ELEMENTARY BIOLOGY

ANIMAL AND HUMAN

[Animal Biology]

BY

JAMES EDWARD PEABODY, A.M.

HEAD OF THE DEPARTMENT OF BIOLOGY, MORRIS HIGH SCHOOL
BRONX, NEW YORK CITY, AUTHOR OF "STUDIES IN PHYSI-
OLOGY" AND "LABORATORY EXERCISES IN ANATOMY
AND PHYSIOLOGY"

AND

ARTHUR ELLSWORTH HUNT, Ph.B.

HEAD OF THE DEPARTMENT OF BIOLOGY, MANUAL TRAINING
HIGH SCHOOL, BROOKLYN, NEW YORK CITY

"The wild life of today is not wholly ours to dispose of as we please. It has been given to us *in trust*. We must account for it to those who come after us and audit our records." — HORNADAY.

New York
THE MACMILLAN COMPANY
1912

All rights reserved

ILLUSTRATED &

PUBLISHED BY

e-KİTAP PROJESİ & CHEAPEST BOOKS

 www.facebook.com/EKitapProjesi

www.cheapestboooks.com

Copyright, 2016

By

e-Kitap Projesi

Istanbul

ISBN:

978-153-941-74-60

Copyright© Printing and publication rights belong to the author's & Publisher's own restriction, using and working. According to the law of intellectual and artistic works, without permission in part or in whole not re-produced or re-published. Welding can be done by showing short excerpts..

COPYRIGHT, 1912,
BY THE MACMILLAN COMPANY.

Set up and electrotyped. Published November, 1912.

YELLOW-BILLED CUCKOOS EATING TENT CATERPILLARS

Photographed from exhibit in Brooklyn Museum of Arts and Sciences, by A. E. Rueff.

TO

THE MEMORY OF

MARTHA FREEMAN GODDARD

WHOSE DEVOTED INSTRUCTION IN BIOLOGY IS A LASTING

INFLUENCE FOR GOOD IN THE LIVES OF HUNDREDS

OF BOYS AND GIRLS AND WHOSE RARE SKILL

IN LEADERSHIP IS AN INSPIRATION TO

EVERY TEACHER WHO KNEW HER

THIS BOOK IS DEDICATED

BY THE AUTHORS

PREFACE

IN the Preface to "Plant Biology" we discussed the general point of view that we believed should be emphasized in a course in elementary biology for students of high school age. We there stated that in our judgment the primary emphasis in the whole course should be placed on the many relations of biology to human welfare. Many of the experiments in that volume, especially those relating to the chemical composition of lifeless and living things, the tests for the food substances, and the principles of osmosis and of respiration, apply equally well in the discussion of animal and human biology.

The method of presentation in "Animal Biology" is somewhat different from that employed in "Plant Biology," for the reason that several widely different types of animals are studied. Limitations of time compel a rigid and somewhat narrow selection of groups for intensive study, and only those functions of each animal are considered which have some relation to human biology, or which have a broad, economic bearing. Thus insects are discussed largely because of their injurious or beneficial effects upon mankind; birds and fishes, because of their economic importance, and because of the great need for their conservation; and one-celled animals because of the light they throw on cellular processes. Certain other somewhat less important topics are considered incidentally; for example, protective resemblance and metamorphosis among insects, and the striking adaptations of structure to function in the bills, feet, and feathers of birds.

The animals suggested for additional study, if time permits, are representative mammals, reptiles, amphibia, arthropods, molluscs, worms, and cœlenterates. In many classes there are students who can work faster than the others, or who are interested in pursuing further their biological studies. Such students may be directed in carrying on some of these studies either in class or outside of school hours. In any case, students are likely to acquire considerable information by reading these textbook descriptions and studying the illustrations.

All the work of the year should lead up to and culminate in human biology. Here, too, however, many important topics must be treated only superficially, or altogether omitted, on account of lack of time. The authors believe that in this, the most important part of the course, practical hygiene should be taught as effectively as possible, and that the necessity for good food, pure air, varied exercise, and sufficient sleep should be continually emphasized. If boys and girls can be led to conform their daily habits to the principles of healthy living, the course in biology will have its highest justification.

In the treatment of Stimulants and Narcotics, the authors have tried to state in simple language the conclusions of experts regarding the effect of tobacco and alcohol, and to present the strongest scientific arguments against the use of these substances which are so injurious to growing youths.

In our judgment there are few, if any, biological topics that are more important in their practical bearing than is that of bacteria. As commonly studied, the disease-producing effects of these organisms are emphasized so much that boys and girls do not appreciate that all the work of the higher organisms depends ultimately upon the activity of these low forms of plant life. In order to bring out this aspect of the work of bacteria, we have called special attention to the structure, physiology, and economic benefit of these organisms. But since so much may be done to prevent disease, we have also considered with some degree of thoroughness the disease-producing effects of several of the pathogenic forms.

No study of human biology should be allowed to leave in the mind of the student the idea that he is merely a chemical engine adapted only for the generation of a certain amount of physical energy. The primary object of all secondary education should be the development of character and efficiency, and the true teacher ought to find opportunity again and again to touch the individual life of the young student. Especially should this be true in the study of biology. Growing boys and girls ought to come to feel, as they have never felt, that they have in their keeping a most complex and wonderful piece of living machinery which can be easily put out of order or even wrecked. But, on the other hand, they should see that if the bodily machine is well cared for, it is capable of splendid work which may help to increase the sum total of human efficiency and happiness.

In the preparation of this volume the authors have received a great many suggestions from the teachers in their own departments and those in other schools. We have been especially fortunate in securing the assistance of experts who have read much of the manuscript and many of the proof sheets. Dr. E. P. Felt, New York State Entomologist, Mr. E. R. Root, author of "A. B. C. of Bee Culture," and Professor Glenn W. Herrick of Cornell University, have given us valuable criticism of the chapter on Insects. Dr. W. T. Hornaday, Director of the New York Zoölogical Park, has read the chapters on Birds and Fishes. To Mr. J. M. Johnson, Head of Department of Biology of the Bushwick High School, we are also indebted for suggestions relating to Birds.

PREFACE

Much of the manuscript of the chapter on Foods received the careful criticism of the late Professor W. O. Atwater. Dr. William H. Park, Director of the Laboratories of the New York Board of Health, and Dr. Thomas Spees Carrington, Secretary of the National Association for the Study and Prevention of Tuberculosis, have given invaluable assistance in the preparation of the chapter on microörganisms. A considerable part of the "Human Biology" was critically read by Dr. F. C. Waite of the Western Reserve Medical School, by Mr. Harold E. Foster of the English Department of the Morris High School, and by the late Miss Martha F. Goddard of the Morris High School, to whose memory these volumes are dedicated. We wish, also, to express our hearty appreciation of the generous permission of Henry Holt & Co. to use some of the material published in Peabody's "Laboratory Exercises in Anatomy and Physiology."

Miss Mabelle Baker, a student in the Morris High School, has contributed the drawings on which her initials appear. To Mr. E. R. Sanborn of the New York Zoölogical Park, and to Mr. A. E. Rueff of the Brooklyn Museum, we are indebted for their skillful photography. The American Museum of Natural History, the Brooklyn Museum, the National Audubon Society, Doubleday, Page & Co., Dodd, Mead & Co., Kny-Scheerer Co., Dr. C. F. Hodge of Clark University, Dr. H. A. Kelly of Johns Hopkins Medical School, Mr. C. W. Beebe of New York Zoölogical Park, and others, have permitted us to make use of illustrative material.

Cost prices for the items on the list of laboratory apparatus and equipment were kindly furnished us by Bausch & Lomb, Kny-Scheerer, and O. T. Louis; from these prices the estimates on pp. 173 to 177 were prepared.

J. E. P.
A. E. H.

October 31, 1912.

CONTENTS

ANIMAL BIOLOGY

		PAGE
PREFACE		vii

CHAPTER
- I. INSECTS 1
    - I. Butterflies and Moths 1
    - II. Grasshoppers and their Relatives 22
    - III. Bees and their Relatives 31
    - IV. Mosquitoes and Flies 43
    - V. (Optional.) Additional Topics on Insects . . 59
- II. BIRDS 62
    - I. Characteristics of Structure 62
    - II. Reproduction and Life History 69
    - III. Methods of Classification 73
    - IV. Importance of Birds to Man 83
    - V. Decrease in Bird Life 91
    - VI. Conservation of Birds 97
- III. FROGS AND THEIR RELATIVES 101
- IV. FISHES 120
    - I. Characteristics of Structure 120
    - II. Adaptations for Nutritive Functions . . 125
    - III. Reproduction and Life History 137
    - IV. Importance of Fishes to Man . . 141
    - V. Conservation of Food Fishes 147
- V. CRAYFISHES AND THEIR RELATIVES 151
- VI. PARAMECIUM AND ITS RELATIVES 164
    - I. Structure and Functions of Paramecium 164
    - II. Structure and Functions of Amœba 170
    - III. Cellular Structure of Higher Animals 172
    - IV. Importance of Protozoa to Man . 173

CHAPTER		PAGE
VII. (OPTIONAL.) ADDITIONAL ANIMAL STUDIES		175
	I. Porifera	175
	II. Cœlenterata	176
	III. Annelida	179
	IV. Mollusca	181
	V. Reptiles	185
	VI. Mammals	187
	VII. Classification of Animals	190
INDEX		197

ELEMENTARY BIOLOGY

ANIMAL AND HUMAN

[Animal Biology]

ANIMAL BIOLOGY

CHAPTER I

INSECTS

I. Butterflies and Moths

1. Insect net. — Since most butterflies and moths are more or less injurious, at least in their caterpillar stage, boys and girls should be taught that they are benefiting their community by catching and killing these insects in a painless manner. For this purpose an insect net and a poison bottle are necessary. An insect net may be made by securing a yard of galvanized iron wire (No. 3), bending it in the form of a ring (thus ⟨), and inserting the two ends of the wire in one end of a light wooden rod about three feet long. To the wire ring should be sewed a bag about two feet deep made of cheesecloth or bobinet (Fig. 1). To catch a butterfly or other insect, wait until it alights, then quickly place over it the opening of the net, holding up the closed end of the net till the insect flies to the top. Now place beneath the insect the open mouth of a poison bottle prepared as follows, and after the insect is in the bottle quickly replace the cover.

2. Poison bottle. — Secure a pint fruit jar or a wide-mouthed bottle fitted with a cover. Into the bottom put a spoonful of more or less pulverized potassium cyanide. Thoroughly mix some plaster of Paris in water and thus make a thin paste. Carefully pour the liquid into the jar until it forms a layer about an inch thick. When this hardens, it covers and holds the cyanide in place, but it is porous enough to allow fumes to escape, which kill most insects in the closed space in a few moments. The bottles are perfectly safe in the hands of pupils. Care should be taken, however, not to handle the cyanide or to breathe in the fumes. The bottle

ANIMAL BIOLOGY

should be kept tightly closed when not in use, and should be distinctly labeled "*Poison Bottle*" (Fig. 2). If the bottle is broken, the pieces of glass and all the contents should be buried in the earth.

3. Preparation of butterflies for study or for collections. — For laboratory study it is desirable to use the largest butterflies obtainable. The work will be carried on to much better advantage if there is at least one mounted specimen for each two pupils. These should be prepared with the wings fully extended, with the legs spread out as in walking, and with the proboscis partly uncoiled. To get the material in this shape place two books about half an inch apart on a soft board; run an insect pin through the thorax of a freshly killed insect, extend the legs and proboscis, then put the body of the insect between the two books, thrusting the tip of the pin into the board beneath. Spread out the fore wings on the book covers so that their hind margins are at right angles to the thorax, pull the hind wings outward into their natural position when at rest, and hold the two pairs in place with pieces of glass till the specimen has dried. Butterfly spreading boards may be bought or made (Fig. 3).

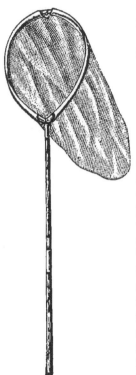

Fig. 1. — Insect net.

Fig. 2. — Poison bottle for killing insects.

Dry specimens may be relaxed by placing a quantity of sand or crumpled paper in a battery jar or other wide-mouthed receptacle that can be tightly covered. Wet the sand or paper thoroughly and then sprinkle over it a little dry sand or cover with blotting-paper.

Put in the dried butterflies about twenty-four hours before they are to be spread, and cover the dish. If the relaxing jar is kept in a warm place, the process will be hastened, but care should be taken not to leave the insects in the moist chamber long enough for mold to grow upon them. It is of course better to mount the butterflies as soon as they are killed.

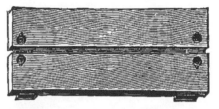

FIG. 3. — Insect spreading board.

4. **Insect boxes.** — A box for displaying a butterfly for class study may be made as described below by any fourteen-year-old boy; these cases will preserve the insects from year to year, thus saving labor as well as insuring good material that pupils can examine from both sides. The boxes may likewise be used as cages for the study of the activities of live grasshoppers, caterpillars, or other insects. After butterflies have been studied they should be transferred to an insect case or other moth-proof box, a piece of cotton soaked in carbon bisulphide should be inserted, and the box kept tightly closed till the butterflies are again needed. " Chiclet " boxes, since they have glass covers, may be used for storing and displaying the insect collections that may be made by pupils. A layer of absorbent cotton over the bottom of the box makes a good background (Fig. 4).

To make the insect boxes, secure from a mill or a local carpenter strips of wood $2\frac{1}{2}$ inches wide and $\frac{1}{2}$ inch thick, with grooves $\frac{1}{8}$ inch wide and $\frac{1}{8}$ inch deep, cut a quarter of an inch from the two margins of one side. About 18 inches will be required for each box. For the sides saw up two pieces each $5\frac{1}{4}$ inches long, and for the ends the

Fig. 4. — Insect collection. (Prepared by Kny-Scheerer Co. Photographed by E. R. Sanborn.)

pieces should be 3¾ inches in length. One of the ends should be planed down to a width of 1¾ inches (the distance between the grooves). Nail the four pieces together and insert in the grooves on each side a cleaned 4 × 5 picture negative, the gelatin of which may be easily removed with hot water. Glue to the center of one of the glasses a piece of cork to hold the insect pin, and fasten a piece of wood to the narrow end by a wire nail, which will prevent the glasses from slipping out but will still allow the box to be opened. The boxes are made more attractive if they are treated with dark oak jap-a-lac or stain (Fig. 5).

Fig. 5. — Insect box.

5. Experiments with living butterflies. — Before trying the feeding experiments, the butterflies should be kept for at least twenty-four hours without food. After a butterfly has fed, it should be placed by itself, since the same insect may be unwilling to eat a second time. Have as many students at a time see the feeding as can well do so; this will save time, and fewer butterflies will be needed. The mourning cloak, monarch, and violet tip butterflies are satisfactory for this experiment. Place the butterfly on a stick or other rough object, and put the tiny drop of honey near it. This may be done in a cage, or under a glass jar, or in the open laboratory.

In the latter case the windows should, of course, be closed, and this should also be done while watching the insect fly. The flying and feeding experiments with insects make excellent home work if the pupils can readily obtain the live material. Children in New York City have caught and kept butterflies for several months, feeding them twice or three times a week.

6. Study of a butterfly. — Laboratory study.

A. *Regions and appendages.*

Examine a butterfly and distinguish (1) the front or *anterior* (Latin, *ante* = before) region called the *head;* (2) the middle region called the *thorax;* and (3) the hind or *posterior* (Latin, *post* = behind) region known as the *abdomen.*
1. Which region is the smallest? Which is the widest? Which region is longest?
2. To which region are the *appendages* (legs and wings) attached?
3. Which region seems to have no appendages?

B. *Organs of the head; feeding.*
1. Observe two long, slender appendages attached to the head; they are called *antennæ* (singular, *antenna*). State the position of the antennæ on the head. Describe the shape of an antenna, stating where it is the thickest (*i.e.* at the *proximal* end, which is next the head, or at the *distal* end, which is farthest from its attachment to the head).
2. Near the base or proximal end of the antennæ find the large *eyes*. State their position on the head, their shape, and their size (as compared with the rest of the head).
3. Demonstration. Take a living or a relaxed specimen of the butterfly, and with the help of a dissecting needle find a coiled structure on the lower or *ventral* surface of the head. It is the *sucking tube* or *proboscis*. Gently uncoil it and describe this feeding organ as to position and appearance.

4. (Optional demonstration or home work.) Place a tiny drop of honey or molasses diluted with water near a butterfly. If the insect does not seem to realize the presence of the sweet substance, touch the proboscis with the needle, or if necessary put the needle into the coil of the proboscis, and gently unroll it.
    a. Describe what you have done to get the animal to eat.
    b. Describe the movements of the proboscis.
    c. What reason do you find for supposing that the butterfly is feeding?
    d. What reason have you for thinking that the proboscis must be hollow?
5. (Optional.) Between the two antennæ, and projecting upward in the anterior region of the head, are two slender structures covered with hair; they are the *labial palps*. In some butterflies the labial palps are inconspicuous. If they show in your specimen, describe them as to their position and appearance.

C. *Organs of the thorax; locomotion.*
1. How many pairs of wings has the butterfly?
2. Describe a wing as to comparative length, breadth, and thickness.
3. Hold a butterfly between your eyes and the light, and study carefully the course of the veins in the two wings on one side. In what region of the wings do the main veins meet?
4. Bend the veins and the connecting membrane in a wing that is given you.
    a. Which is the more rigid?
    b. What, then, is one use of the veins?
5. Take a small piece of the wing of a butterfly that is given you and rub the surface with your finger tip.
    a. Describe what you have done, and state how the substance on your finger compares in color with the color of the part of the wing before it was rubbed.

b. (Optional.) Shake some of the powder from a wing upon a glass slide and examine it with a low power of the compound microscope. The bodies that you see are called *scales*. At one end of each scale you should find a tiny stem by which the scale was attached to the wing, and at the other end usually one or more notches. Describe the shape of the scales that you are studying, and make a sketch of one of them much enlarged.

6. (Optional home work.) Watch a butterfly in the field as it moves the wings in the act of flying.

a. Will the downward stroke of the wings tend to lower or to raise the body?

b. What effect will the upward stroke of the wings tend to have?

c. In which of these two directions, therefore, must the butterfly strike the harder and more quickly in order to raise the body in the air?

d. Since the weight of the body tends to bring the animal to the ground, in which direction must the insect strike with the greater force in order to keep itself at a given level in the air?

7. Some butterflies have a tiny pair of front legs that are usually folded against the thorax; so that you need to look very carefully before deciding as to the number of legs present.

a. How many pairs of legs has this insect?

b. Are the legs long and slender or short and thick?

c. Is each leg all one piece or is it jointed as in the human body?

d. Examine the lower end of a leg and state how the foot is adapted for clinging to flowers.

D. Make a drawing, natural size, of the upper or *dorsal* surface of a butterfly. Label antennæ, eyes, proboscis, head, thorax, abdomen, wings, principal veins of one wing.

7. General characteristics of butterflies. — All butterflies, as we shall see later, are constructed on much the same general plan as that of other insects; *i.e.* their bodies are divided into three regions, head, thorax, and abdomen; on the head are two antennæ and a pair of large eyes; on the thorax are two pairs of wings and three pairs of jointed legs; and the abdomen is composed of a number of parts called rings or *segments* (Fig. 6).

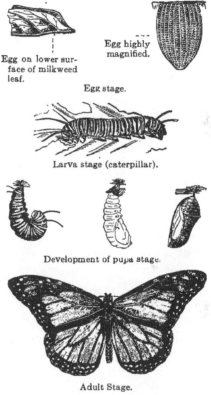

Egg on lower surface of milkweed leaf.

Egg highly magnified.

Egg stage.

Larva stage (caterpillar).

Development of pupa stage.

Adult Stage.

FIG. 6. — Life history of monarch butterfly. (Weed.)

8. Wings and their scales. — While this general plan of structure is common to all insects, there are certain marked peculiarities that enable one readily to recognize a butterfly.

For instance, although other insects have two pairs of wings, no others have these organs so beautifully colored and relatively large. This color of the wings is due (we proved in **6**, *C*. 5) to tiny bodies called *scales*. If the wing of a butterfly is rubbed, the color comes off and the wing at that point loses its color. To the unaided eye this colored substance from the wing appears to have no definite form; in fact, it looks like the pollen from flowers. An examination with the compound microscope, however, shows that each of these tiny bodies has a definite shape (Fig. 7). Each scale has at one end a tiny stem, but in other respects they vary considerably in form.

The scales are attached in the following manner. In the membrane of the wing are openings into which fit the stems of the scales. The latter are arranged in rows and overlap something like the shingles on a roof

FIG. 7. — Scales from wing of a butterfly.

(Fig. 8). In spite of this arrangement it is evident that the scales are not firmly attached, since the slightest touch is sufficient to dislodge many of them. Rough handling was not apparently planned for in the construction of these insects. The presence of these scales on the wings of butterflies and of their near relatives, the moths, is so characteristic that these insects have been called the *Lepidoptera* (Greek, *lépido* = scale + *ptéra* = wings). Not only are scales found on the wings but, in the shape of hairs, they form a fuzzy growth over the surface of the whole body.

FIG. 8. — Piece of the wing of a butterfly with scales. (Coleman.)

9. Proboscis. — Another marked characteristic of butterflies and moths is the sucking tube, or *proboscis*. While the proboscis seems to be a single structure, in reality it is composed of two slender appendages, each having a groove on its inner surface; so that, when the two parts are brought together, they form a tube through which the butterfly sucks nectar from flowers. When the proboscis is not in use, the butterfly rolls it into a tight coil underneath the head (Fig. 9).

10. Legs. — The legs of a butterfly are not very strong, since they are relatively so long and slender. This is perhaps the reason why these insects seldom use them for walking. They are, however, very useful in clinging to flowers.

Fig. 9. — Head of butterfly. (Coleman.)

The two curved claws on the tip of each foot show clearly the means by which the animals are able to hold on to the plants on which they usually alight.

11. Reproduction and life history of butterflies. — As in the reproduction of plants, the development of the butterfly begins with a special cell known as an *egg-cell*. These egg-cells are formed in the body of the female insect. When these egg-cells have been fertilized by *sperm-cells* from the male butterfly, which correspond to sperm-cells of the pollen grains (P. B.[1], 91), the eggs are deposited on the under side of the leaves of plants on which the young can feed (Fig. 6). These egg-cells divide and subdivide, till at last a many-celled organism is developed that is commonly called a " worm," but that is more correctly known as a *caterpillar* (Fig. 6).

The tiny caterpillar emerges from the covering of the egg and begins to feed upon the leaf. As it feeds it grows, and

[1] P. B. = " Elementary Plant Biology," by the authors of this book.

from time to time sheds or *molts* the more or less hardened skin that covers the whole insect. At last, after several molts, the caterpillar reaches its full size and then stops eating. At no time in the growth of the caterpillar would one be likely to mistake it for a butterfly (Fig. 6). It has no wings, no antennæ, and instead of a proboscis one finds a pair of strong jaws with which it eats leaves. The distinction between thorax and abdomen is not at all clear, and at first sight it seems to have more legs than a butterfly. The three front legs are really jointed, but they are so short and thick that there seems to be no resemblance between them and those of a butterfly. The other pairs of legs, varying in number, are not jointed structures, and hence are not really legs at all.

The mature caterpillar now attaches itself to some object and, after molting once more, usually assumes quite a different shape from that of the caterpillar, and forms about itself a hardened skin within which a marvellous transformation occurs (Fig. 6). The long, coiled tube takes the place of the jaws as a feeding organ, and long, slender, knobbed antennæ appear on the head; two pairs of beautifully colored wings develop on the thorax, as well as the three pairs of slender, jointed legs; and at last the fully developed butterfly breaks through the covering that held it and flies away.

It is evident, then, that a butterfly passes through several fairly distinct stages. First we may distinguish the *egg stage*, then the caterpillar or *larva stage*, which is followed by the transformation stage in which it is called a *pupa*. The pupa of a butterfly is often called a *chrysalis* (Greek, *chrysos* = gold) on account of the golden spots of color on many pupa cases. Lastly we have the fully developed or *adult insect* that emerges from the pupa stage.

12. Distinguishing characteristics of moths. — The moths and butterflies belong to the same order of insects; that is, the scaly winged insects. But there are some characteristics in which these two kinds of insects differ. For instance, moths when at rest fold the wings horizontally (Fig. 11), while butterflies fold them vertically, that is, erect (Fig. 10). The wings, too, of moths are not usually as brilliantly colored. Most moths fly at night, while butterflies are day-flyers. The body of moths is usually relatively broader than that of butterflies. Moth antennæ are of various shapes, often like a feather, but never knobbed.

In general, the life history of moths is very much the same as that of butterflies, but the larvæ of many moths spin a more or less silky mass of threads about themselves, as is the case with the silkworm caterpillar (Fig. 16), and this outside covering of the pupa stage is known as the *cocoon*.

13. Economic importance of butterflies and moths. — The larvæ of both butterflies and moths are voracious feeders, as any one knows who has had any experience with caterpillars. In fact, they may be called animated feeding machines, since the animal must not only provide for its own growth, but must also store up enough food to form the new parts such as the wings and the legs. Not all larvæ of butterflies and moths are considered harmful, however, since some of them are not prolific enough to have any serious effect upon vegetation, which is the source of food of most caterpillars. This is true of many of the butterfly larvæ and of some moth larvæ. Then, too, some of the larvæ feed on plants that are not useful to man. This is true of the larva of the monarch butterfly (Fig. 6), which feeds upon leaves of the milkweed. The adult butterflies and moths of course are not capable of doing any harm since, when they eat anything at all, they most commonly suck the nectar of flowers. When the flowers are visited in this way, they are very likely to be cross-pollinated and thus are benefited instead of injured. But in general the moths and butterflies play but little part in the very important process of cross-pollination of flowers, most of this work being done, as we shall soon learn, by the bees. The following are a few of the injurious forms of butterfly and moth larvæ.

14. Cabbage butterfly. — This is one of the few forms of butterfly larvæ that are of sufficient economic importance to be worthy of mention. Any one who has been near a cabbage patch will remember to have seen many rather small white butterflies (Fig. 10) hovering about among the cabbages. These are the cabbage butterflies depositing their eggs on the under side of the leaves. The small green caterpillars that develop from the eggs very soon show what they can do in the way of eating. The ragged appearance of the young leaves is a warning to the gardener to "get busy" if he desires a crop. The caterpillars do most harm when the cabbages are young, since these plants may be so injured as to be unable to form heads. The caterpillars are often killed by sprinkling with a mixture of Paris green and arsenate of lead in water (47). This mixture should not be used, however, after the heads begin to form, on account of the possibility of the poison collecting between the leaves of the head, with consequent danger to the consumer.

FIG. 10. — Life history of cabbage butterfly. (Coleman.)

15. Tussock moth. — The caterpillars of the tussock moth attack our shade trees. Where they are unchecked, they will practically

INSECTS 15

strip the trees of their leaves. The female moth is wingless (Fig. 11). When she emerges from her cocoon, she lays a mass of eggs upon the

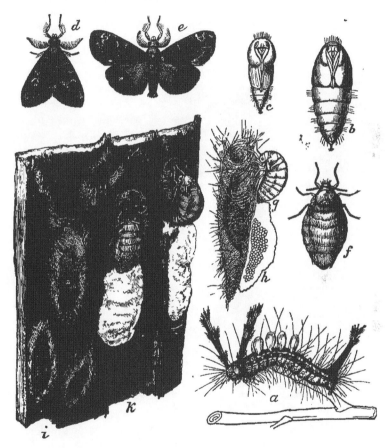

Fig. 11.— Life history of tussock moth. (Osborn.)

outer surface of the cocoon and secretes about them a white foamy mass which hardens (Fig. 11). If this occurs in the autumn, the eggs remain during the winter, and the following spring hatch out. The young caterpillars attack the leaves of the tree on which they have hatched out, or if the cocoon was placed elsewhere, they crawl up the nearest tree and start business at once. They are great travelers, and this is the way they spread through a neighborhood, since, as already mentioned, the female cannot fly. To capture these insects one may place a band of cotton batting around the trunk of each of the trees one wishes to protect. The larvæ do not usually crawl over this but will, if mature, proceed to pupate underneath the band. All pupæ and egg masses should be collected (Fig. 12) and burned. This is about as much as the individual can do. Where a spraying apparatus is available the trees should be sprayed with lead arsenate, thus killing all the caterpillars. This caterpillar is rather handsome as caterpillars go, having a bright red head and a series of yellow tufts of hair on the dorsal part of the body (Fig. 11).

FIG. 12. — Morris High School boys removing 63,020 eggs of tussock moth from four trees on school grounds. Work directed by Paul B. Mann. (Photographed by Lewis Enowitz.)

16. Gypsy moth and brown tail moth. — The gypsy moth (Fig. 13) was brought into Massachusetts from Europe in 1869 in connection with scientific experiments. Some of these specimens acci-

INSECTS 17

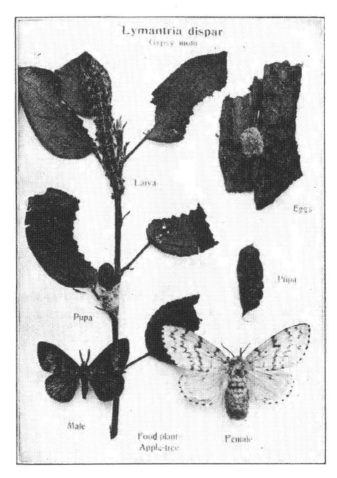

FIG. 13. — Life history of gypsy moth. (Prepared by Kny-Scheerer Co. Photographed by E. R. Sanborn, N. Y. Zoölogical Park.)

dentally escaped and gradually increased until the damage to fruit, forest, and shade trees caused by the larvæ was so evident that property owners had to call upon the state to aid in their extermina-

18　　　　ANIMAL BIOLOGY

tion. Nearly one million dollars was expended during a period of ten years. At the end of this time the number of the insects was so reduced that it was impossible to convince taxpayers of the necessity for further appropriations to complete the extermination. Since then the gypsy moths have spread over the whole state of Massachusetts and into the adjoining states.

The larvæ of another moth, the brown tail, has likewise caused great damage in the New England states. The New York State Department of Education is sending out colored pictures of the life history of both of these insects with the following statement regarding them. " Warning — Take Notice. There is grave danger of both of these dangerous pests being brought into New York State. They have destroyed thousands of trees in Massachusetts, and they will do the same in New York unless checked. All are hereby urged to become familiar with the general appearance and work of these two insects, and to report anything suspicious to the State Entomologist, Albany, N.Y., sending specimens if possible. Abundant hairy caterpillars an inch to two inches long on or in the vicinity of defoliated trees should lead to investigation."

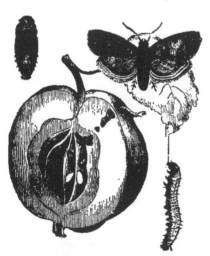

FIG. 14. — Life history of codling moth.
(U. S. Dept. of Agriculture.)

17. Codling moth. — Every one has eaten into apples that have been injured by the " apple worm," which is the larva of the codling moth (Fig. 14). The damage to the fruit crop from this insect in New York State alone is estimated at three million dollars each year. According to Professor Hodge (" Nature Study and Life ") the cod_ ling moth "was early imported from Europe and is now at home wherever fruit is cultivated in this country and Canada, causing a loss of from 25 to 75 per cent of the apple crop, as well as that of many other fruits. In the heavy bearing years the wormy apples fall off and are discarded, but the great number of apples serves to rear enormous numbers of the worms, and, according to my observations and experience, in the off years, when apples would be valuable, the worms take the whole crop.

"The larvæ change to pupæ in May, emerge as moths in late May or June, and lay their eggs for the first brood in June. The larvæ generally crawl into the calyx cup of the young apples and eat their way to the core, complete their growth in about three weeks, commonly eat their way out through the side of the apple, and either spin to the ground and crawl to the trunk of the tree or crawl down the branches and make their cocoons under the bark again. This occurs with the greater number early in July. This habit affords one of the most vulnerable points of attack. To trap practically all the codling moths in an orchard it is only necessary to scrape all loose bark off from the trees and fasten around the trunks a band of burlap or heavy paper. Remove the bands and collect all larvæ once a week during July." The practice of most commercial growers at the present time, however, is to depend very largely or entirely on spraying with a poison (*e.g.* arsenate of lead, **47**). One application, even, a week or ten days after the blossoms fall, if thorough, will frequently give 95 per cent to 98 per cent of sound fruit.[1]

18. Clothes moths. — ".The little buff-colored clothes moths (Fig. 15) sometimes seen flitting about rooms, attracted to lamps at night, or dislodged from infested garments or portières, are themselves harmless enough, for their mouth parts are rudimentary, and no food whatever is taken in the winged state. The destruction occasioned by these pests is, therefore, limited entirely to the feeding or larval stage. The killing of the moths by the aggrieved

---

[1] The authors are indebted to Mr. E. P. Felt, state entomologist of New York, for this and several other suggestions relating to insects.

housekeeper, while usually based on the wrong inference that they are actually engaged in eating her woolens, is, nevertheless, a most valuable proceeding, because it checks, in so much, the multiplication of the species which is the sole duty of the adult insect.

"There is no easy method of preventing the damage done by clothes moths, and to maintain the integrity of woolens or other materials which they are likely to attack demands constant vigilance, with frequent inspection and treatment. In general, they are liable to affect injuriously only articles which are put away and left undisturbed for some little time. . . . Agitation, such as beating and shaking, or brushing, and exposure to air and sunlight, are old remedies and still among the best at command. Various repellants, such as tobacco, camphor, naphthalene cones or balls, and cedar chips or sprigs, have a certain value if the garments are not already stocked with eggs or larvæ. Furs and such garments may be stored in boxes or trunks which have been lined with the heavy tar paper used in buildings. New papering should be given to such receptacles every year or two."[1]

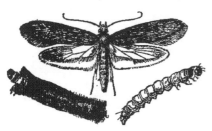

FIG. 15.—Life history of clothes moth. (U. S. Dept. of Agriculture.)

19. Silkworms. — One species of moth, the silkworm (Fig. 16), is of great economic importance to man. The larva of this insect feeds upon the leaves of the mulberry tree, and after reaching maturity it spins a cocoon, requiring about three days for its completion. The silk is obtained by heating the cocoon in ovens to kill the pupa, and then by reeling off the silk and spinning it into threads. "For many hundreds of years the cultivation of the silkworm was confined to Asiatic countries. It seems to have been an industry in

[1] Circular No. 36, Second Series, United States Department of Agriculture.

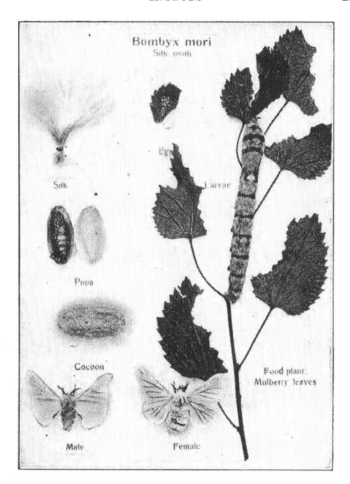

Fig. 16. — Life history of silkworm moth.· (Prepared by Kny-Scheerer Co. Photographed by E. R. Sanborn, N. Y. Zoölogical Park.)

China as early as 2600 B.C., and was not introduced into Europe until 530 A.D. After the latter date the culture rapidly increased, and soon became prominent in Turkey, Italy, and Greece, and has held its own in those countries, becoming of great importance in Italy. . . . Japan to-day produces a very considerable proportion of the world's supply of raw silk. Thus of the $41,000,000 spent by the United States for raw silk in 1902, more than $20,000,000 went to Japan."[1] Many attempts have been made to introduce this industry into the United States, but the experiments thus far made have been rather unsuccessful.

II. GRASSHOPPERS AND THEIR RELATIVES

20. **Study of the grasshopper.** — Laboratory study.

A. *Regions and appendages.* — Examine a grasshopper and distinguish the three regions of the body proper: (1) the front or *anterior* region called the *head;* (2) the middle region called the *thorax;* and (3) the hind or *posterior* region known as the *abdomen.* (The anterior region of the thorax is covered by a cape or collar.)
   1. Which region is the smallest? Which is the widest? Which region is the longest?
   2. Which region has legs and wings attached to it?
   3. Which region is made up of a number of similar rings or *segments?*

B. *Organs of the head; feeding.*
   1. Notice two long, slender feelers on the head. They are known as *antennæ* (singular, *antenna*). State the position of the antennæ on the head and describe their shape.
   2. Describe the shape and position of the large eyes. State their relative size compared to that of the head.
   3. (Optional.) Cut off with a sharp knife a thin slice from the outer surface of one of the large eyes. Remove all the soft, dark material from the inside. Place the

[1] Bailey's Cyclopedia of Agriculture, Vol. III, p. 640.

cleaned piece on a glass slide and examine the outer (convex) surface with the low power of the compound microscope. Look for the boundary lines of many several-sided areas. Each of these areas is called a *facet*. Each facet is the covering of one of the parts of which the *compound eye* is composed.
   a. Describe the preparation of the slide for examination.
   b. Describe the shape of each of the facets, and make an outline drawing of three of them, much enlarged, to show the way in which they fit together.
4. (Optional.) With the aid of a magnifier look for a tiny eye in the middle of the front part of the head. There is a similar eye between each compound eye and the antenna of the same side. These eyes are *simple eyes*. Describe the simple eyes as to location, number, and relative size.

5. Find the *upper lip* (*labrum*) on the lower anterior part of the head. Describe its location and shape.
6. (Demonstration.) Raise the upper lip of a large grasshopper and find the *jaws* or *mandibles* beneath it. With a dissecting needle gently pry the jaws a little way apart. Do the jaws move from side to side or up and down?

7. (Optional.) Find the *lower lip* on the under side of the head, *i.e.* next to the thorax. It is divided vertically into two equal parts. Attached to either side are two tiny, jointed structures called *labial palps*.
   a. Describe the location of the lower lip (*labium*).
   b. Describe the position and appearance of the labial palps.
8. (Optional demonstration.) Between the jaws and the lower lip of a large specimen find a pair of appendages each of which is made up of three parts that are joined together at the base: (1) on the outside is a several-jointed feeler or *maxillary palp;* (2) next is a spoon-shaped body; and (3) a curved and sharp-pointed

body. It will be necessary to pull sideways on the mouth parts to see this inner part. These three parts form one appendage called the *maxilla* (plural *maxillæ*), or helping jaws.

When you have found these three parts of a maxilla, describe them.

9. (Demonstration or home work.) Place several grasshoppers in a cage or a glass jar with moistened leaves of clover, grass, or lettuce. If these insects refuse to eat, try others till you find some that will eat.
    a. Describe the movements of the head and also the movements of the mouth parts while the grasshopper is eating.
    b. Which mouth parts must do most of the biting of the leaf? Give reason.

10. (Optional.) Make a drawing, at least four times natural size (× 4) of the face view of a grasshopper. Label antenna, compound eye, simple eye, upper lip.

C. *Organs of the thorax; locomotion.*

1. How many legs has a grasshopper? Which pair is the largest?
2. Make a sketch (× 4) to show the following parts of one of the hind legs: (1) a large segment nearest to the thorax, the *thigh* or *femur;* (2) the next segment to the femur, the *tibia;* (3) the part that rests on the ground when the insect walks, the *foot* or *tarsus*. Use a magnifier to see the several segments in the tarsus, the little claws at the tip end. and a little pad between the claws. Label femur, tibia, segments of the tarsus, claws, pads.
3. (Optional.) Make a sketch (× 4) of one of the smaller legs to show the size and shape of the parts. Use the same labels as in the drawing of the hind leg.

INSECTS

4. Get a grasshopper to climb up a stick or piece of grass.
    a. Tell what you have done and observed.
    b. How is the insect able to cling to the stick?
5. (Demonstration or home work.) Place a lively grasshopper in a clear space on the floor or in a cage. Get it to jump enough times to determine the following points: —
    a. What is the position of the parts of the hind legs when the animal is ready to leap?
    b. What is the position of the parts of the hind leg the instant the insect lands?
    c. What does the grasshopper do to get ready for another jump?
    d. What movement throws the insect into the air? Is this movement made slowly or quickly?
    e. In what respects are the hind legs better fitted for jumping than are the two other pairs?
    f. What seems to be the use of the smaller pairs of legs when the insect lands on plants?
6. Move the outer wings sideways and forwards at right angles to the body so as to expose the under pair. Spread out or unfold the under wings. (It is an advantage to mount the specimens on cork and pin the wings in the position named above.)
    a. Which pair of wings is better fitted for flying? Why?
    b. How are the outer wings fitted to protect the under wings?
    c. (Optional.) Draw ($\times$ 2) the outline of a front wing and of a hind wing, and sketch in the principal veins. Label front wing, hind wing, veins.

D. *Organs of the abdomen; breathing.*

1. You will observe that each of the rings or segments of the abdomen is composed of an upper or *dorsal* half and an under or *ventral* half. Make a sketch ($\times$ 4) of a side view of four or five segments of the abdomen to show the structures mentioned above.

2. Secure an active grasshopper, put it in a live cage, and watch the movements of the upper and lower halves of the abdominal segments. Describe what you have observed.
3. In each segment, except those at the tip of the abdomen, there are two *breathing pores* or *spiracles*, one on each side. With the aid of a magnifier look for these breathing pores near the lower margin of the dorsal half of each segment. When you have found the spiracles in four or five segments, show them in your sketch (1, above), and label breathing pores or spiracles.
4. The spiracles lead into tiny elastic *breathing tubes* or *tracheæ* (singular *trachea*) which extend throughout all parts of the body of the insect even into the wings. The veins that you can see in the wings contain these minute tubes. Describe the tracheæ and state their extent and their connection with the spiracles (Fig. 17).
5. The tracheæ have an elastic material in their walls, so that when they have been compressed, they will spring back to their former shape and size as soon as the pressure is removed. Describe, now, the structure of one of the air tubes, and state what action this structure makes possible.

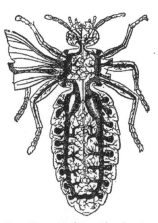

Fig. 17. — Air tubes (tracheæ) of an insect.

6. When the under or ventral half of the abdomen moves up into the dorsal half —
   a. Will the diameter of the abdomen be increased or decreased?

  b. Will the air tubes be made larger or smaller? Why?
  c. Will the air now rush into the air tubes or out of them? Why?
 7. If the upper and lower halves of the abdomen now move apart —
  a. Will the diameter of the abdomen be increased or decreased?
  b. How will this movement affect the size of the tracheæ? Why?
  c. Will the air now move into the tracheæ or out through the spiracles? Why?

21. Characteristics of grasshoppers. — After studying two or three insects, the student will see that they all resemble the grasshopper (1) in having three regions of the body (head, thorax, and abdomen), (2) in possessing as appendages one pair of antennæ, one pair of compound eyes, two pairs of wings, and three pairs of legs, and (3) in having an abdomen made up of a number of rings or segments.

Fig. 18. — Mouth parts of a cockroach. (Parker and Haswell.)

The most distinguishing characteristics of the grasshopper and its relatives are found in its mouth parts and wings. Grasshoppers have biting mouth parts throughout their life. These consist of (1) an upper lip that is notched, (2) a pair of horny jaws, or *mandibles*, (3) a pair of rather complicated helping jaws or *maxillæ*, and (4) a lower lip. The two lips move up and down while the two pairs of jaws move from side to side. All these structures are well adapted for holding and biting off leaves of grass or other plants, and this seems to be the main business of this insect.

Grasshoppers, too, are admirably provided with organs of locomotion. In fact, they derive their name from the extraordinary feats of jumping, which they accomplish largely by their long and muscular hind legs. If a boy could jump twenty times the length of his legs, that is, a distance of 50 feet, he would make an athletic record corresponding to that of the common red-legged locust. For the hind legs of an ordinary specimen of this insect are about 2 inches long, and they frequently leap 4 feet. The wings are also of great assistance in enabling the animal to secure its food or to escape its enemies. Flight is accomplished by the help of the hind pair only, and when these are not in use, they are folded like a fan beneath the outer pair.

22. Life history of the grasshopper. — The male grasshopper may be easily distinguished by the rounded tip of the abdomen; the abdomen of the female, on the other hand, has at its posterior extremity four movable parts which constitute the egg-laying organ or *ovipositor* (Fig. 19). The eggs are produced within the body of the female insect. Before these eggs can develop, however, each must be fertilized by a sperm-cell produced by the male grasshopper, just as an egg-cell of a plant must be fertilized by the sperm-nucleus of a pollen grain (**P. B., 91**). After the process of fertilization has taken place, the female grasshopper (usually in the fall of the year) burrows a hole in the ground by alternately bringing together, pushing into the earth, and then spreading apart, the four projections that make up the

Fig. 19. — Grasshopper laying eggs. (Riley, U. S. Dept. of Agriculture.)

ovipositor (Fig. 19). From 20 to 40 small, banana-shaped eggs are then laid in the bottom of the hole. In the spring each egg hatches into a tiny grasshopper, which much resembles the adult, except that it has no wings and its head is relatively large in comparison with the rest of the animal. The insect begins at once to feed and grow, but since its whole exterior is hard and resistant, growth can only take place after this outer covering has been split and the insect has crawled out. This process is known as *molting*, and takes place five or six times during the life history of the animal. The insect then forms a new and larger coat. At each molt the wings become more fully developed, until at the last molt the adult insect is produced (Fig. 20). Hence, in the life history of the grasshopper there are three more or less distinct stages: (1) the egg, (2) the developing insect, which is known as the *nymph*, and (3) the adult grasshopper. This succession of changes in a life history is known as *metamorphosis* (Greek, *meta* = one after another + *morphos* = form). But, because in the development of the grasshopper these changes are not so striking as those that occur in the life history of the butterfly (11), the metamorphosis of the grasshopper is said to be *incomplete*. It is better, however, to refer to it as a *direct metamorphosis*, that of the butterfly being known as an *indirect metamorphosis*. After reaching the adult stage and depositing eggs, the adult insects die. Only a few of the immature grasshoppers survive the winter, and these are the grasshoppers that are seen early in Spring.

FIG. 20. — Stages in life history of a grasshopper.

23. Economic importance of grasshoppers. — Our laboratory study of a grasshopper's mouth parts and our observations of its methods of feeding have shown that these insects resemble caterpillars, first, in having biting mouth parts (Fig. 18), and second, in being voracious eaters. Hence, as we should expect, a large number of grasshoppers in a given area would mean a considerable destruction of plant life. Many " plagues of locusts " (for grasshoppers are more correctly known as locusts) have been recorded in history. One of the first is that recorded in the Bible, which occurred before the departure or " Exodus " of the Children of Israel from Egypt. " And they (the locusts) did eat of every herb of the land, and all the fruit of the trees and there remained not any green thing in the trees, or in the herbs of the field throughout all the land of Egypt." (Ex. x. 15.)

In our own country during the years 1866 to 1876 there were several plagues of locusts in the grain-producing states of the West, notably in Kansas and Nebraska. The Rocky Mountain grasshoppers during these years migrated in such numbers that the sky was darkened during their flight, and the result of their devastation was as serious as that described in Exodus. According to one authority this species of insect destroyed $200,000,000 of crops in the western states in the space of four years. No great migrations have occurred since 1876.

Locusts have been used as food, and even at the present day they are commonly eaten by the Arabians. In the Bible, it is related of John the Baptist, that while preaching in the wilderness "he did eat of locusts and wild honey."

FIG. 21. — Four walking sticks on a branch. (Coleman.)

INSECTS

24. Relatives of the grasshopper. — Other insects that have structure, habits, and life history similar to those of the grasshopper are the crickets, cockroaches, katydids, and walking sticks.

The cockroaches are more commonly known in New York City as "Croton bugs" from the fact that they frequent places close to water pipes through which Croton water is carried. They are very fast runners, as any one knows who has tried to catch them, and their bodies are so thin that they can easily hide away in narrow cracks. Their sharp jaws enable them to feed upon dried bread and other hard food (Fig. 18).

Katydids and walking sticks are striking examples of *protective resemblance;* that is, they resemble their surroundings in form or color so closely that they may secure protection from their enemies by this means (Fig. 21).

III. BEES AND THEIR RELATIVES

25. A study of the bumblebee. — (Laboratory study.)

A. *General survey.*
1. Give the names of the regions that you find in the body of the bee. (See **20**, *A*.)
2. State the number and situation of the antennæ. (See **20**, *B*.)
3. How many compound eyes are present, and where are they situated? (See **20**, *B*, 3.)

4. (Optional.) With the help of a magnifier look for the simple eyes on the top of the head and between the compound eyes. How many simple eyes are there, and what is their color? (See **20**, *B*, 4.)

5. Examine the legs and state —
    a. Their number, and the region of the body to which they are attached.
    b. The relative size of the different pairs.
    c. Their adaptations (by structure) for walking.

6. Examine the wings and state —
    a. Their number, and the region of the body to which they are attached.
    b. Their characteristics of texture.
    c. Their adaptations for flying.
7. Is the abdomen segmented or not?

B. *Food-getting organs.*
1. If the mouth parts do not project from the lower part of the head, you should find them bent backward beneath the head and thorax. Use the dissecting needle to straighten them out. Carefully separate these mouth parts and count them.
    a. How many mouth parts do you find?
    b. Describe the general shape of all these parts.
    c. How are the mouth parts fitted to enable the bee to get nectar from flowers?

2. (Optional.) Spread the mouth parts on some white blotting paper and stick pins into the blotting paper so as to keep the parts from coming together. Use the magnifier to distinguish the following parts: —
    a. The central, longest part, the *tongue*. (It has hairs on its surface.) The tongue springs from a broader body, the *lower lip*.
    b. Two shorter parts on either side of the tongue, springing also from the lower lip, and called *labial palps* because they are believed to correspond to the jointed bodies of that name attached to the lower lip of the grasshopper and other insects. (See **20**, *B*, 7.)
    c. Two broader parts springing from a point farther back than the labial palps and supposed to correspond to the helping jaws of the grasshopper, and hence called *maxillæ*. (See **20**, *B*, 8.)
    Draw a front view of the outline of the head and of these five mouth parts (× 4). Label each part.

3. (Optional.) The bee also has a pair of small *mandibles*. They are attached to the head below the compound eyes. They extend forward and are often crossed underneath the lower lip. Separate them carefully with the dissecting needle. When the bee uses them, it bends the other mouth parts back out of the way.
   a. Are the mandibles hard or soft?
   b. Describe their color and shape.
      Note. — The honeybee uses the mandibles in forming the wax cells of the comb, and also at times as organs of defense.

4. Examine the hind leg and find the following parts: —
   a. A fairly prominent segment nearest the thorax, the *femur;*
   b. A segment larger than the femur and just below it, the *tibia;*
   c. A broad segment below the tibia, the basal part of the foot or *tarsus;*
   d. The remainder of the tarsus or foot consisting of four tiny segments with hooks on the end segment.
      Make a drawing of one of the hind legs ($\times 4$) to show all these parts in outline and label femur, tibia, basal part of tarsus, hooks, tarsus.
5. Examine the outer surface of the tibia with a magnifier, noticing several rows of hairs around the margin. The portion of the tibia that faces outward, together with the hairs, is called the *pollen basket.*
   Locate the pollen basket and show how it is adapted for holding pollen.

26. History of beekeeping. — "It is abundantly evident from the records of the remote past that beekeeping has always been a favorite occupation with civilized nations. Egypt, Babylon, Assyria, Palestine, Greece, Rome, and Carthage all had their beekeepers. In the days of Aristotle (in Greece) there are said to have existed two or three hundred treatises on bees, so that, then as now, beekeeping was a favorite topic with authors. More books have appeared on bees and bee-culture than have ever been published about any domestic animal, not excepting the horse or the dog." [1]

Yet from the earliest times until the middle of the last century there was little improvement in the method of keeping bees. They were allowed to build their combs in hollow trunks of trees or in hives so constructed that it was impossible to control in any way the work of the bees (Fig. 22). In 1852, however, Rev. Lorenzo Langstroth of Philadelphia invented a hive with movable frames, and his invention wholly revolutionized the beekeeping industry. Practically all modern hives throughout the world are constructed on the plan that he introduced, which is essentially as follows. In a rectangular box are suspended eight to ten movable frames, in each of which the bees build their comb, store honey, and develop their young; for this reason this part of the hive is known as the *brood chamber*. (One of these frames, covered with bees is shown in Fig. 23.) As the season advances, the beekeeper places above the brood chamber successive *supers* (Latin, *super* = above), each supplied with little boxes (Fig. 23) which when filled with honeycomb usually weigh about a pound. It is this excess of stored honey that is commonly offered for sale.

FIG. 22.—Old type of beehive. (From International Encyclopedia. Dodd, Mead & Co., N. Y.)

[1] Cyclopedia of American Agriculture, Vol. III, p. 278.

27. Characteristics and functions of the queen and the drones. — Honeybees, though smaller than bumblebees, resemble them in their general plan of structure; that is, both kinds of insects have a head, thorax, and abdomen, all

FIG. 23. — Modern type of beehive.

more or less covered with hair, and on the thorax are two pairs of membranous wings and three pairs of jointed legs. In every colony of bees there is, except at rare intervals, only one queen. The queen-bee (Fig. 24) can be readily distinguished from all the other individuals in the hive by her long, slender abdomen (Fig. 24). It is her sole business to deposit an egg in each of the various wax cells of the brood chamber. Queens have been known to lay 3000 eggs in a single day, and since a queen may live as long as five years, she may lay over 1,000,000 eggs during a lifetime. The queen is therefore the mother of all the bees in a colony.

The distinguishing characteristics of drone or male bees (Fig. 24) are their broad abdomens, the absence of a sting, and their very large, compound eyes, which nearly meet on the top of their heads. In numbers they vary at different

FIG. 24. — Drone, queen, and worker bee.

times of the year, but during the summer there are usually 400 to 800 in a hive.

We learned in our study of reproduction in plants that egg-cells will not develop into seeds unless they are fertilized by sperm-cells of pollen grains. Now in a beehive, an egg will never develop into a queen-bee or a worker unless it likewise is fertilized by a sperm-cell. The drones or male bees supply these necessary sperm-cells. From the unfertilized eggs, which a queen may lay, develop only drone bees. In this respect these egg-cells of bees are strikingly different from those of plants and of most animals.

It is clear from the foregoing account that the queen and drones carry on the *reproductive functions* of the colony, for they are specially adapted to increase the number of bees in a hive. To the workers, on the other hand, as we shall now see, belong most of the *nutritive functions* of the colony.

INSECTS

28. Characteristics of worker bees. — While the workers are smaller than either the queen or the drones, they are by far the most numerous, there being as many as 50,000 in a good colony in midsummer. In shape they resemble the queen, as one would expect, since they are undeveloped female bees. As was the case with the bumblebee, their mouth parts are very complicated, consisting of a central tongue and two other pairs of appendages, all of which form a hollow tube for sucking up the nectar of flowers (Fig. 25). Above the tongue is a pair of horny jaws that move from side to side, which the bees use mainly for comb building. On the tibia of each hind leg of a worker bee is likewise a fringe of stiff hairs, which, together with the concave outer surface of the tibia, forms a pollen basket similar to that of the bumblebee. In this the insect gathers a mass of pollen which may easily be seen when the workers are returning to the hive (Fig. 26).

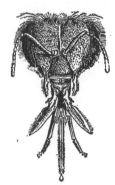

FIG. 25. — Mouth parts of a bee.

FIG. 26. — Hind leg of bee with pollen, inner surface.

29. Comb building. — All the work of comb manufacture is carried on by the worker bees, and when one studies this process carefully, it is found to be one of the greatest marvels of animal activity. The cells of the comb are built out horizontally from each side of a central partition in a brood frame or of a super box. To save the bees' time and to insure even comb, beekeepers usually insert in the frames or honey boxes thin sheets of wax "foundation" on which the bases of the cells have been impressed by machinery. Upon this the workers build the comb outward. But without this assistance from man the comb cells are usually remarkably regular and show the greatest economy in the use of wax. The cross section of each cell is a hexagon, and so these compartments fit together without any spaces between them as would occur if the cells were cylinders. (See Fig. 27.) This hexagonal shape also permits a single partition wall to serve for two adjacent cells, and it is evident that this shape of cell more closely fits the body of the bee than would a four-sided cell. The worker bees build two different sizes of cells in the comb. Most of the cells average about twenty-five to a square inch, and in these the fertilized eggs are laid, which, as we have said, develop into workers. The cells in which unfertilized eggs are deposited are somewhat larger. These form the so-called *drone comb*.

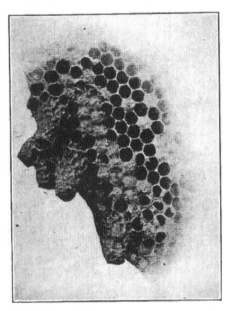

Fig. 27. — Worker cells and queen cells. (From "A, B, C of Bee Culture." A. I. and E. R. Root.)

The wax from which the comb is produced oozes out from certain glands on the ventral surface of the abdomen of the workers. When producing the wax the bees hang motionless inside the hive for several days, each holding to the bees above. They have al-

INSECTS

ready gorged themselves with honey, and it is estimated that from seven to fifteen pounds of honey are required to produce one pound of wax. As the little plates of wax are formed, they are seized by a bee and carried with its mandibles or under its "chin" to the comb where the building is going on. Here the wax is pressed against one of the walls.

30. Honey making. — While studying flowers we learned that they secrete a sweet liquid known as nectar. It is this that the workers use for honey manufacture. The bee inserts into the blossom its sucking tongue and pumps up the nectar into a sac known as the *honey stomach* (Fig. 28). Here a kind of digestion takes place whereby the nectar is changed to honey. If the worker bee is hungry, it opens a little trapdoor and allows the honey and pollen to pass into the true stomach. But since the insect usually makes more honey than it can use, when it returns to the hive it squeezes its tiny honey stomach and deposits the surplus in the cells of the comb. This honey, when first made, contains a good deal of water; it would therefore take up too much room in the comb and it would be more likely to run out from the horizontal cells. Hence, some of the workers fan with their wings and evaporate the surplus water. When the cells are completely filled, they are capped over with wax.

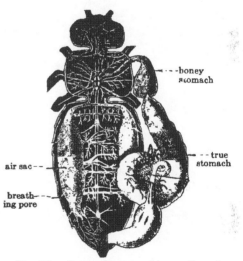

FIG. 28. — Internal organs of bee. (Lang.)

31. Other duties of worker bees. — Bees, we have also learned (28), bring in large quantities of pollen packed in the pollen baskets of the hind legs, and in gathering pollen a considerable amount clings to the head and other parts of the body. Worker bees also bring in from the buds of trees a brown, gummy substance called bee glue or *própolis* which they use to close up crevices in the inside of the hive. In most hives, too, certain bees seem to be detailed to act as soldiers to keep out individuals from another swarm or other marauders which might raid their stores of food. During the busy summer season a worker usually lives only a month or two.

Certainly enough has been said to convince any one that a bee colony is a wonderful social community, organized more completely, so far as division of labor is concerned, than many a human community. Is it a monarchy ruled by the queen, or a democracy controlled by the workers? The latter is more probably the case. Yet we can hardly imagine how the thousands of individuals can work together in such a helter-skelter way and accomplish such wondrous results.[1]

32. Life history of the honeybee. — The *eggs* of the bee are tiny white objects, shaped more or less like a banana. A single egg is fastened by the queen mother at the bottom of each cell in the brood comb (Fig. 29). At the end of three days the egg hatches into a minute footless *grub* or *larva* (Fig. 29) which is fed for the first few days on rich food, produced in the stomach of the

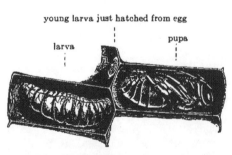

Fig. 29. — Stages in life history of honeybee. (Cheshire).

[1] For interesting descriptions of the work carried on in a beehive see "A, B, C of Bee Culture," by A. I. and E. R. Root.

workers that are acting as nurses. The grubs are then fed with a mixture of pollen and honey, and at the end of six days after hatching they are supplied with enough of this mixture to last during the rest of the larva stage, and the cells are then capped over with wax by the workers. There the developing bees pass through the third or *pupa stage* (Fig. 29), and at the end of twelve days bite their way out of their nursery cells and take their share in the busy toil of the hive.

Drones, we have said, develop in somewhat larger cells than worker bees. When the colony wishes to produce a queen, the workers build a cell about as large as the end-joint of one's little finger (Fig. 27), and as soon as the egg is hatched they stuff the little grub throughout the larval stage with what is called " royal jelly," never giving it the undigested pollen mixture that is supplied to the grubs of workers or drones.

33. Swarming. — We come now to one of the most interesting events in the story of bee colonies. If several queens emerge from their cells at the same time, they attack each other in a royal battle, for it is said that a queen never uses her sting except against a rival. When the conflict is over, the victorious queen becomes the mother of the hive. For in the meantime the former queen, surrounded by half the drones and workers, has left the old hive, abdicating in her daughter's favor. After emerging from their old home, the swarm of bees thus formed alights on a neighboring tree, clinging to each other in a solid mass. It is then comparatively easy for a beekeeper to shake the insects from the limb into a new hive, and if the queen is secured, the swarm will usually begin work at once in their new home (Fig. 30). If, however, the bees are not captured, scouts go out to search for a hollow tree; and when satisfactory quarters are found, the whole swarm follows their guides, and build their comb in the home thus secured.

34. Economic importance of bees. — In our study of flowers we referred frequently to the necessity of the visits of bees to insure cross-pollination. Indeed, Professor Hodge says ("Nature Study and Life") that for all practical purposes

Fig. 30. — A swarm of bees on a limb. (Lyons.)

so far as man is concerned, the honeybee is sufficient for this purpose (with the exception of securing a red clover crop, which requires the help of the bumblebee). In years to come we may be sure that the most successful fruit farmers will also keep bees.

It is estimated that the annual production of honey and wax in the United States amounts to between twenty and thirty millions of dollars, and if scientific management were to be introduced more widely, this output could be raised to fifty million dollars a year without additional investment. Almost any one who is interested can keep bees. During a single season the swarm in the observation hive on the fourth floor of the Morris High School in New York City produced fifty-six pounds of honey in the super boxes, besides laying by in the brood chamber a sufficient supply for their winter support.

35. Relatives of the bees. — Wasps and hornets belong to the same order of insects as the bees, and resemble them more or less closely in structure. Some kinds of wasps build paper comb from wood which they chew up with their jaws. Ants, insects with which every one is familiar, are likewise classed with the bees and wasps, and the social communities that they form are marvelous in the degree to which they carry division of labor. In some ant colonies in addition to the workers there are soldiers and slaves.

IV. Mosquitoes and Flies

36. Life history of the common inland or house mosquito. — The eggs of the common house mosquito are laid by the female in little rafts that float on the surface of stagnant water. These egg masses look like flecks of black soot, but when examined with a hand lens each is found to consist of 200 to 400 cartridge-shaped eggs standing on end (Fig. 31). If the weather is warm and other conditions are favorable, the eggs hatch within a day into tiny mosquito larvæ, which are known as "wrigglers" from their characteristic motion in the water.

In the second stage in its life history, which usually lasts about a week, the mosquito larva feeds on the microscopic plants and animals that abound in all stagnant water, and grows rapidly. Just as was the case with the butterfly and moth caterpillars, this rapid growth necessitates the frequent shedding or *molting* of the outer covering of the larva and the formation of a new and larger coat. Hence, in water

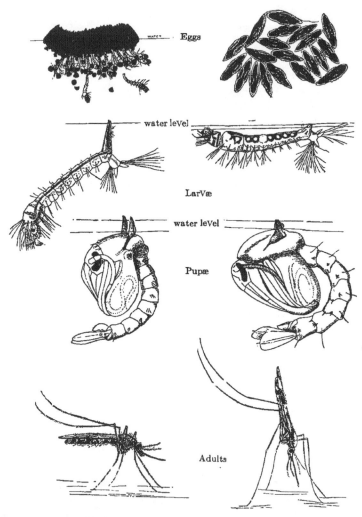

Fig. 31. — Life history of house mosquito (Culex).   Fig. 32. — Life history of malaria mosquito (Anopheles).
(Howard, U. S. Dept. of Agriculture.)

where mosquitoes breed, one finds countless "suits of cast-off clothing" which would fit all stages of the young wrigglers.

The mosquito larva has a well-developed head, a thorax, and a jointed or *segmented* abdomen, but legs and wings are wanting. The most striking characteristic of this stage of the mosquito is the breathing tube that projects diagonally from the hind end of the abdomen. For while the mosquito larva lives in the water, it is obliged to swim to the surface at short intervals to get its necessary supply of air. It then hangs diagonally with the tip of its breathing tube projecting through the surface film into the air above (Fig. 31). This habit frequently proves its undoing, as we shall see when we come to discuss the methods of mosquito extermination.

After attaining its full growth as a larva, the insect enters the third or pupa stage (Fig. 31). " The pupa," says Miss Mitchell in her " Mosquito Life," " is the form intermediate between the larva and the adult. Unlike most pupæ, those of the mosquito are very active, but like other pupæ, they do not eat. They are about the shape of fat commas, floating quietly at the surface or bobbing crazily downward at the least alarm to hide at the bottom, propelled by backward flips of the abdomen The creature no longer breathes through a single tube on the eighth segment of the abdomen but by means of a pair of tubes on the back of the thorax." During this stage the insect develops its sucking mouth parts, its long, slender legs, and its two delicate wings, and all these organs may be seen through the transparent outer coat, which is composed of a substance known as *chitin*.

At the final molt the mosquito leaves its pupal case in the water and flies into the air, an adult mosquito. If it hatches during the spring, summer, or early autumn, it usually lives no more than a week or two; but many of the female mosquitoes that develop late in the autumn seek out a protected spot in which to spend the winter, and thus are ready in the spring to perpetuate the species by laying eggs in the stagnant pools formed by early rains.

All that the mosquito needs, therefore, in order to develop its offspring from egg to adult stage is a bit of water that will remain relatively undisturbed for about two weeks. Hence, old tomato cans, bits of crockery, and other receptacles carelessly left in many a back yard, furnish breeding places for all kinds of mosquitoes.

Truth compels us to remark, in passing, that the male mosquito is a decent sort of fellow, keeping close to his breeding place and feeding on plant juices or eating nothing at all during his brief existence in the adult stage. It is the lady mosquito that torments us by singing her piercing song and piercing our suffering skins. But as in most other sufferings that we endure, the fault is largely our own. At least we can secure immunity if as communities we but persist in applying the simple methods of extermination outlined in **42**.

37. Life history of malaria-transmitting mosquitoes.—The mosquito we have just described, while a nuisance wherever found, does not, so far as is known, cause disease. There are, however, two kinds of mosquitoes that are not only a nuisance but a menace to life and health wherever they are found; namely, those that transmit malaria and yellow fever. The first of these is the *Anópheles* mosquito, commonly known as the "malaria mosquito," for as we shall soon see, malaria cannot be transmitted from one human being to another except through the agency of this species of insect. The eggs of the Anopheles mosquito are larger than those of the house mosquito and are laid singly, not in masses (Fig. 32). In the larva stage, likewise, the two insects may be easily distinguished from the fact that the "malaria wriggler," while breathing, lies horizontally just beneath the surface of the water, while the other species hangs downward, with only the tip of the breathing tube projecting to the water level (Figs. 31 and 32).

In Figs. 31 and 32 the characteristic position of the adults of the two species is shown. While the body of the house mosquito is usually parallel to the surface on which it alights, that of the malaria-transmitting insect is sharply tilted away from the surface.

38. Occurrence of malaria. — The story of the discovery that a kind of mosquito known as the Anopheles mosquito is the only means, as far as we now know, by which malaria may be transmitted from one individual to another, is one of the most wonderful in all the history of biology. In a guide leaflet on "The Malaria Mosquito" published by the American Museum of Natural History, New York City,[1] the author, B. E. Dahlgren, writes as follows: —

"It was early observed that 'malaria' was apt to be prevalent during the damp and rainy seasons, and that it occurred principally in exactly such places as are now known to furnish ideal breeding grounds for the malaria mosquito. That new cases of malaria appeared at the time of year when the Malaria Mosquito abounded, was also recorded long before it was suspected that the insect was in any way con-

[1] Every one who visits the American Museum should study carefully the wonderful set of models that show on a big scale the various stages in the life history of the mosquito. These models are pictured in the bulletin referred to above, which may be obtained from the librarian of the Museum for fifteen cents.

nected with the malady; and one of the old medical writers mentions as a characteristic of malaria seasons that 'gnats and flies are apt to be abundant.' . . .

"Malaria was formerly considered to be a form of ague due to foul air, whence its name, which literally means 'bad air.' It was attributed to a sort of 'miasma.' Its true nature did not become known till 1880, when Laveran, a French military surgeon, working, at the time, in Algeria, discovered the malarial parasite in human blood." Major Ross, an English officer in India, later proved the presence of the parasite in the body of the mosquito.

39. Transmission of malaria. — Investigation has shown that the parts of the world where Anopheles abound are the eastern half of the United States and a large part of Europe, together with many regions of the tropics. It is a well-known fact that these are the regions, too, in which malaria is very abundant, and this is the first line of proof that the Anopheles mosquito is always responsible for the transmission of malaria.

Even more conclusive were the experiments of four investigators who spent the fever season in the dreaded malaria district of the Roman Campagna. They built for themselves a carefully screened house in which they remained from sunset to sunrise, and this was the only precaution that they observed. In the daytime they went freely among those who were stricken with the fever, they allowed themselves to be soaked with the falling rains, and at night the air from the swamps came freely into their sleeping quarters. But while hundreds of malaria cases were all about them, not one of the four contracted the disease. Hence, to escape malaria, one has only to make sure that Mrs. Anopheles is prevented from injecting her billful of malaria germs — and this she does only during night time, " loving darkness, rather than light, because her deeds are evil."

In the year 1890 Dr. Manson and Dr. Warren, two physicians in London, allowed themselves to be bitten by Anopheles mosquitoes that had previously bitten malaria patients in Italy. In eighteen days both developed malarial fever and in the blood of both, malaria organisms were found, although previous to this infection from the mosquito neither had suffered in any way from the disease.

40. Life history of the malaria parasite. — But yet more wonderful proof that the mosquito transmits malaria has been furnished by the microscopes of biologists. The discovery of the malarial parasite by Laveran in 1880 has already been referred to. This resembles in its form and activities a single-celled animal known as the Amœba (124). When this organism of malaria is present in human blood, it bores its way into a red corpuscle (H. B., 6), feeds upon the contents of this blood cell, and grows at the expense of the corpuscle until the parasite occupies nearly all the space inside it (Fig. 33). The malaria parasite then divides into a number (6–16) of daughter parasites, which rupture the red corpuscle in which they have been developing, and escape into the liquid part of the blood, thus causing the chills so characteristic of malaria. Each new parasite then attacks a new corpuscle and at the end of two or three days produces six to sixteen new spores, and so the organisms multiply.

Now here comes the relation of the mosquito to malaria. For when the female Anopheles bites a person having malaria, she is likely to suck up blood that contains malaria organisms in a certain stage of development. These reach the insect's stomach, where they pass through a stage known as fertilization. In the stomach of a single mosquito as many as five hundred of these fertilized cells have been counted. Each cell then becomes pointed at one end, bores its way through the wall of the mosquito's stomach, and in fifteen to twenty days produces relatively large swellings on the outer surface, in which are thousands of needle-shaped malaria spores (Fig. 33).

At length these spores escape through the outer wall of the mosquito's stomach and many of them find their way to the salivary glands. And so when the infected mosquito bites another person, these parasites are injected with the saliva, and if the conditions are favorable in the blood of the new victim, the spores straightway attack the red corpuscles, and a new case of malaria is the result.

For the treatment of malaria quinine is the most effective drug known at present. It should be taken in the quantity and at the times prescribed by the physician.

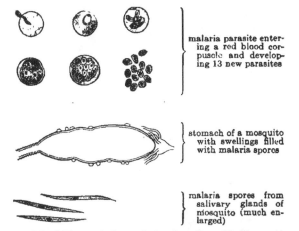

Fig. 33. — Life history of the malaria parasite. (Dahlgren, American Museum Natural History.)

41. Transmission of yellow fever. — The proof that malaria can only be transmitted from one human being to another was largely the work of the biologists of England, France, and Italy. The discovery that another kind of mosquito (*Stegomyia*) is responsible for the transmission of the parasite that causes yellow fever is due almost wholly to the splendid achievements of the Yellow Fever Commission appointed by President McKinley. In June, 1900, this commission of five, headed by Dr. Walter Reed (Fig. 34), began its epoch-making experiments in the Island of Cuba, and within six

INSECTS 51

months these men demonstrated conclusively that this plague disease of the tropics and of our southern states can, so far as we know, be communicated only through the agency of the Stegomyia mosquito.

This commission, believing in the mosquito theory, at once began experiments to demonstrate its truth. One of the members, Dr.

FIG. 34. — Dr. Walter Reed.

Lazear (Fig. 35), permitted a mosquito to bite him; a few days later he contracted the disease and died. The inscription on a tablet erected in his memory reads as follows: "With more than the courage and devotion of the soldier, he risked and lost his life to show how a fearful pestilence is communicated and how its ravages may be prevented."

When Dr. Reed called for volunteers from among the soldiers, the first to respond " was a young private from Ohio, named John R.

FIG. 35.—Dr. Jesse Lazear.

Kissinger (Fig. 36), who volunteered for the service, to use his own words, ' solely in the interest of humanity and the cause of science.' When it became known among the troops that subjects were needed for experimental purposes, Kissinger, in company with another young private named John J. Moran, also from Ohio, volunteered their services. Dr. Reed talked the matter over with them, explaining fully the danger and suffering involved in the experiment should it be successful, and then, seeing they were determined, he stated that a definite money compensation would be made them. Both young men declined to accept it, making it, indeed, their sole stipulation that they should receive no pecuniary reward, whereupon Major Reed touched his cap, saying respectfully, 'Gentlemen, I salute you.' Reed's own words in his published account of the experiment on Kissinger are: ' In my opinion this exhibition of moral courage has never been surpassed in the annals of the Army of the United States.' "[1]

The object of one of the first experiments was to determine whether or not yellow fever could be contracted from clothing worn by yellow fever patients. A small building was constructed the windows and doors of which were carefully screened. Into this were brought chests of clothing that had been taken from the beds of patients who had been sick and in some cases had died of yellow fever. Three brave men entered the building, unpacked the boxes, and for twenty nights slept in close contact with the soiled clothing. "To pass

FIG. 36. — John R. Kissinger, U.S.A.

twenty nights in a small, ill-ventilated room, with a temperature over ninety, in close contact with the most loathsome articles of dress and furniture, in an atmosphere fetid from their presence, is an act of heroism which ought to command our highest admiration and our lasting gratitude."[2] In spite, however, of their unwholesome surroundings, none of the men contracted yellow

[1] From "Walter Reed and Yellow Fever," by Dr. H. A. Kelly Doubleday, Page & Co.   [2] "Walter Reed and Yellow Fever."

fever, and so it was proved for all time that this disease cannot be communicated by means of anything that comes from the body of yellow fever patients.

Dr. Reed now sought to prove that the *Stegomyia* mosquito was the means by which the disease was transmitted from one person to another. A second building, the same size as the first, was erected, the room was divided by a wire screen, and all the doors and windows were carefully screened (Fig. 37). Into one of the rooms a number of mosquitoes that had bitten yellow fever patients were freed and a

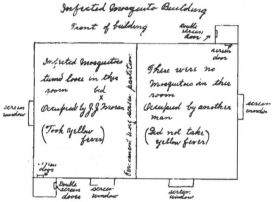

FIG. 37. — Plan of infected mosquito building. (Drawn for the authors by John R. Kissinger.)

few minutes later John Moran, an Ohio soldier, entered and allowed these mosquitoes to bite him. "On Christmas morning (1900) at 11 A.M. this brave lad was stricken with yellow fever and had a sharp attack which he bore without a murmur." On the other side of the screen were three soldiers who were protected from mosquitoes; and these men remained in perfect health. This experiment proved conclusively that yellow fever is transmitted by the Stegomyia mosquito.

42. Extermination of mosquitoes. — Now all the suffering from malaria and yellow fever is entirely unnecessary

INSECTS 55

if communities will but take the trouble to eradicate all breeding places of mosquitoes. Since the mosquito, during its development in the water, comes frequently to the surface to secure air for breathing, a thin film of oil spread over the surface of the water in which they are breeding is a sure means of killing them. But the kerosene treatment is at best but a temporary means of ridding a community of mos-

FIG. 38.—Staten Island marshes before drainage.

quitoes. The oil has to be renewed every two or three weeks, especially after rains, to make sure that a continuous film covers the surface. Hence, wherever possible, pools should be drained, and one has but to read Dr. Doty's account of his marvelous success in abating the mosquito nuisance on Staten Island (see New York State Journal of Medicine, May, 1908) to be convinced that this method is effective (Figs. 38 and 39). Every householder should coöperate by cleaning up his own back yard and by covering cisterns and wells with the finest meshed netting; for the insistent mosquito has been known to make its way through ordinary wire netting. Fish and dragon flies are also important helps to man, since these animals devour great numbers of larvæ, pupæ, and adult insects; yet at best they can hardly be counted as efficient means of ridding swamps of mosquito pests.

Anopheles, Culex, and Stegomyia lay their eggs in the same

FIG. 39. — Staten Island marshes after drainage.

kind of stagnant pools, and the proper filling, draining, or screening of these pools or their treatment with kerosene, will destroy the one as well as the other. When this is accomplished, malaria and yellow fever, as has been conclusively demonstrated by the work of Americans on Staten Island, New Orleans, Cuba, and Panama, will practically disappear.

INSECTS

43. Habits and life history of the house fly. — It has been clearly proved that the common house fly is a frequent cause of disease; especially is this true in the transmission of typhoid fever and the intestinal diseases to which the deaths of so many young children are due. Practically all parts of the body of a fly are covered with hairs (Fig. 41), especially the mouth parts and feet. Each foot, also, has sticky pads (Fig. 40) which enable the fly to cling to the walls and ceilings. In the adult stage the flies feed upon filth of all sorts, and if they alight on the excretions of typhoid patients, they are very likely to carry on their feet and mouth parts the germs of the disease, and so when they come into the house, they may infect milk and other food. Flies also carry germs of other diseases such as cholera, dysentery, and tuberculosis.

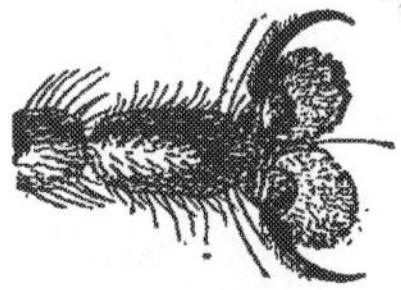

Fig. 40. — Foot of fly, showing hairs and pads.

The most common breeding place of house flies is in piles of horse manure. Here the female fly lays about 120 eggs (Fig. 41) which hatch within a few hours into tiny white footless grubs. These feed for about five days upon the manure and grow rapidly, molting twice within that time. The larva now changes into a pupa, and at the end of another five days the adult fly emerges from the brown pupa case. Egg laying begins almost at once, and as each adult female fly lays 120 eggs, it has been estimated that a single fly may have 5,598,720,000 descendants in a single season if each fly were to deposit but one batch of eggs. In reality, however, a fly deposits four batches in a season. Hence it is very important to catch and kill flies at the very beginning of each season.

44. Extermination of house fly. — Because of the danger of disease transmission every housekeeper should do her best to screen her house and so keep flies away from food, for one never knows where these insects have been crawling or what disease germs may be clinging to their feet. City authorities

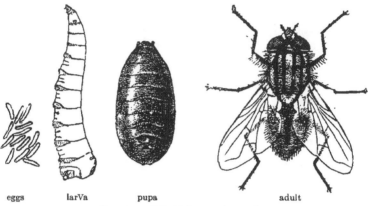

FIG. 41. — Life history of house fly.

should see that all street refuse and garbage are removed before flies of any kind can lay their eggs therein. All persons

FIG. 42. — Life history of potato beetle. Identify eggs, larvæ, pupa and adult.

responsible for horse stables should make sure that the manure is thrown into screened pits and sprinkled with chloride of lime at least once a week.

Another method of dealing with the problem is that suggested by Professor C. J. Hodge of Clark University, Worcester, Mass. It is that of letting the flies catch themselves. He has devised a simple and inexpensive flytrap, which is easily attached to any garbage can (or to a window screen); or it may be baited with bits of fish or other food. The flies are attracted by the odors of the garbage or food bait, and when caught may be killed with boiling water. If the various suggestions are followed, even farmhouses, as experience has shown, may be rendered practically free from the filthy and dangerous house fly.

V. Additional Topics on Insects

45. Field and library study of other insects. — (Optional.) Study as many of the following insects as time allows, consulting Sanderson's "Insect Pests of Farm, Garden, and Orchard," Hodge's "Nature Study and Life," National and State Circulars and Bulletins, articles in Encyclopedias or other reference books. Emphasize especially the habits, life history, and economic importance of each of the following insects: Colorado potato beetle (Fig. 42), cut worms, army worms, San José scale (Fig. 43), tent caterpillar, chinch

Fig. 43. — San José scale insects on pear. Above, single scale enlarged. (Howard.)

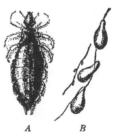

Fig. 44. — A, human louse; B, eggs attached to hair.

bug, cockroaches, plant lice, human lice (Fig. 44), bedbugs (Fig. 45), carpet beetles, lady bugs, scavenger beetles, ichneumon fly.

46. Annual loss due to insect pests of the United States.[1] — "In no country in the world do insects impose a heavier tax on farm products than in the United States. The losses resulting from the depredations of insects on all the plant products of the soil, both in their growing and in their stored state, together with depredations on live stock, exceed the entire expenditures of the national government, including the pension roll and the maintenance of the army and the navy." This loss for the year 1904 was estimated at $795,100,000, and this does not include the expense involved in applying insecticides.

Fig. 45. — Bedbug.

Product	Value	Percentage of Loss	Amount of Loss
Cereals	$2,000,000,000	10	$200,000,000
Hay	530,000,000	10	53,000,000
Cotton	600,000,000	10	60,000,000
Tobacco	53,000,000	10	5,300,000
Truck crops	265,000,000	20	53,000,000
Sugar	50,000,000	10	5,000,000
Fruits	135,000,000	20	27,000,000
Farm forests	110,000,000	10	11,000,000
Miscellaneous crops	58,000,000	10	5,800,000
Animal products	1,750,000,000	10	175,000,000
Total	$5,551,000,000		$595,100,000
Natural forests and forest products			100,000,000
Products in storage			100,000,000
Grand total			$795,100,000

[1] C. L. Marlitt in Year Book of United States Department of Agriculture, 1904. The figures are regarded as conservative estimates.

INSECTS

47. Insecticides.[1] — In our laboratory studies we have found that there are two kinds of insects, namely those with biting mouth parts and those with sucking mouth parts. Entirely different treatment is necessary in dealing with insect pests of these two types. Insects with biting mouth parts may usually be killed by thoroughly spraying the parts of a plant upon which they feed with a mixture made in the following proportions : —

	1 Gal. Mixture	50 Gal. Mixture
Arsenate of lead (poison) . .	$\frac{2}{5}$ oz. (1 teaspoonful)	$2\frac{1}{2}$ lb.
Water	1 gal.	50 gal.

Insects with sucking mouth parts, on the other hand, must be treated with a spraying mixture which will actually touch their bodies. Some of the " contact insecticides " for this purpose are whale oil soap (one pound to five gallons of water), kerosene emulsion, and " black-leaf-40." The last named is preferable for killing plant lice and it is mixed with soap as follows : —

	1 Gal. Mixture	50 Gal. Mixture
Black-leaf-40 (40 per cent nicotine)	$\frac{1}{8}$ oz. (= 1 spoonful)	$\frac{1}{2}$ lb.
Ivory or laundry soap . . .	$\frac{1}{8}$ oz. (size of two yeast cakes)	2 lb.
Water	1 gal.	50 gal.

[1] The authors are indebted to Professor Glenn W. Herrick of Cornell University for these formulas.

CHAPTER II

BIRDS

48. Study of a bird. — (Optional home work.)

So far as possible the following study should be made from a robin, sparrow, chicken, or other living bird, and the observations should be supplemented by an examination of stuffed specimens, charts, or pictures.

A. *Regions.*

In all animals that have internal bony skeletons as do birds, at least two of the following regions may be distinguished; namely, a head, a neck, a trunk, and a tail.

Which of the four regions named above can you distinguish in the bird that you are studying?

B. *Head.*

1. Describe the general shape of the *beak* (or *bill*), stating whether it is relatively long and slender, or short and thick. State, also, whether the tip of the beak is straight or curved.
2. On what part of the head are the *eyes* located?

   In the eyes of a bird the following parts are visible: a central *pupil*, and around this a colored region known as the *iris*. State the location and describe the color of each of these parts in the eye of the bird that you are studying.
3. In front of the eyes find two openings, the *nostrils*. Locate the nostrils with reference to the beak and the eyes.
4. Make a drawing twice natural size of a side view of the head to show the beak, eye, and nostril. Label each part shown.

5. Watch a chicken, canary, sparrow, or other bird while it is eating and drinking, and describe the movements that the bird makes in these acts.

C. *Organs of locomotion.*
   1. What is the position of the wings when they are not in use?
   2. Note and describe the movements of the wings when the bird is flying.
   3. The only part of the leg that is visible in most birds is the foot, the upper parts being covered with feathers.
      a. How many toes do you find pointing forward and how many backward? (Be sure to name the kind of bird on which this observation is made.)
      b. Make a drawing to show the foot and the toes with the claws at the end of each. Label toes and claws.
   4. Watch several kind of birds (*e.g.* robins, sparrows, chickens, starlings), and state whether each of these kinds of birds walks or hops.

49. What is a bird? — Birds, like fishes, frogs, and man, belong to the group of animals that have a backbone, and hence are known as *vertebrates*. It is never difficult, however, to distinguish birds from other vertebrates, since every bird has *wings* either developed or undeveloped (Fig. 46) and a covering of *feathers*. Birds, too, maintain a body *temperature* that is higher than that of any other group of animals. The temperature in man, for instance, is normally about $98\frac{1}{2}°$ F., whereas no bird, so far as we know, has a temperature less than 100° F., and even 111° F. is known to be the temperature of some of the sparrows and warblers. Hence, we may define *a bird as a warm-blooded vertebrate, having wings and a body covering of feathers and usually able to fly.*

Even a casual examination will show that a bird has a *head, neck,* and *trunk,* and two pairs of appendages, namely, the *wings* and *legs*. With the exception of the feet, practically the whole of the animal is covered with feathers (Fig. 46).

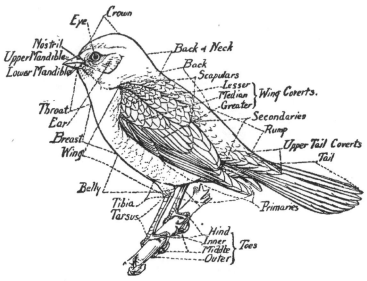

Fig. 46. — External structure of a bird.

50. Head. — A closer study of a bird shows that from the front part of the head projects a horny structure known as the beak or bill. "Tie a man's hands and arms tightly behind his back, stand him on his feet, and tell him that he must hereafter find and prepare his food, build his house, defend himself from his enemies and perform all the business of life in such a position, and what a pitiable object he would present! Yet this is not unlike what birds have to do. Almost every form of vegetable and animal life is used as food by one or another of the species. Birds have most intricately built homes, and their methods of defense are to be numbered by the score; the care of their delicate plumage alone would seem to necessitate many and varied instruments: yet all this is made possible, and chiefly executed, by one small portion of the bird — its bill or beak." [1]

While the size and shape of the bill varies greatly in different kinds of birds, it always consists of two parts (*mandibles*) (Fig. 46), which correspond in position to the upper and lower jaws of man. When the bill is opened, a careful examination shows that a bird has no teeth. Some of the birds that lived ages ago, however, had well-developed teeth in their jaws, as is well shown in (Fig. 47) which is a picture of a bird skeleton restored from bones found in the rocks of western Kansas.

Fig. 47. — Skeleton of a fossil bird.

Near the base of the bill on either side, one can usually see an opening; these openings are the *nostrils*. On the sides of the head are the two *eyes*, and since they bulge out somewhat, the bird is afforded a wide range of vision. If the feathers below and behind the eye are pushed aside, an opening into the *ear* may be seen; this may be made out easily in the head of a chicken.

51. Wings. — In Figure 48 are shown the bones that compose the wing of an ostrich and the arm of a man, and on comparing the two one sees a striking resemblance. In both, the upper arm has a single bone, while in the forearm there

[1] Beebe, "The Bird."

are two bones. In the hand region, though the differences are more striking, the general plan of the two is the same. Unlike the bones of the human skeleton those of most birds are hollow and filled with air.

Any one who has eaten a chicken's wing knows that the bones are covered by muscles; these enable the bird to fold and unfold the parts of the wing, much as the human arm is stretched out or doubled up. On the bird's body are other powerful muscles, which cause the wing as a whole to make the upward and downward strokes in flight.

Still another wonderful adaptation of the wing for flight is evident in the arrangement and structure of the feathers (Fig. 49). The feathers fit over each other in such a way[1] that in the downward and backward stroke of the wing a continuous surface is struck against the air, and this propels the bird upward and forward. In the up-

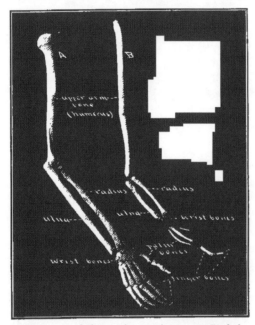

FIG. 48. — *A*, skeleton of arm of a man; *B*, skeleton of wing of an ostrich. (A. E. Rueff.)

[1] Before assigning these paragraphs the structure of a feather and the arrangement of the feathers on the wing of some bird (*e.g.* a chicken) should be demonstrated to the class.

BIRDS 67

ward wing stroke, on the other hand, the resistance of the air is diminished since the feathers are separated more or

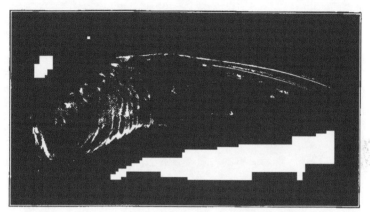

FIG. 49. — Wing of Tern. (Photographed by E. R. Sanborn, N. Y. Zoölogical Park.)

less like the slats of a Venetian blind, thus allowing the air to pass between them.

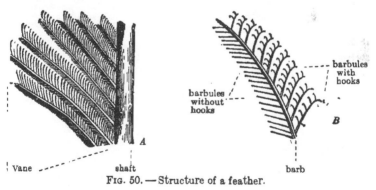

FIG. 50. — Structure of a feather.

An examination of a single feather [1] shows that it consists in the first place of a *shaft* running through its length (Fig.

[1] See footnote, p. 66.

50, *A*). On the sides of the shaft are the two flat surfaces which make up the *vane*. This vane is composed of slender parts called *barbs* that may be easily separated from each other, or when separated may be readily united, because of little hooks (**Fig. 50,** *B*). This the bird does when it smooths or "preens" its feathers.

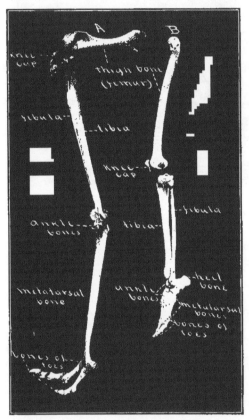

Fig. 51.—*A*, skeleton of leg of an ostrich; *B*, skeleton of leg of a man. (E. R. Sanborn.)

52. Legs.— On comparing the arm of man with the wing of a bird we found that they were similar in structure, and the same is likewise true of the leg and foot. While the thigh of a bird is much shorter proportionately than is that of man (Fig. 51), both have but a single bone. Below the knee of the bird is the shank or "drumstick" which consists of a long bone extending to the ankle, and beside it is a slender bone attached only at the upper end. This region in the leg of man is likewise composed of a relatively thick shin bone, on the outer side of which is a thin bone extending down to the ankle.

BIRDS 69

The ankle region of a bird is the joint half-way up the leg (Fig. 51, *A*). What is commonly regarded as the bird's foot consists often of three toes that point forward, and one that extends backward. Ordinarily the parts of the leg below the ankle are covered with scales, and the tips of the toes are provided with claws.

53. Study of a hen's egg. — (Optional home work.)

Secure the egg of a hen or other domestic bird, and study it as follows: —

1. Describe the difference in the shape and size of the two ends of the egg.
2. Carefully crack the shell at the larger end and remove the pieces of shell.
    a. State what you have done and describe the *membrane* that lines the shell.
    b. Carefully cut this membrane and note that the liquid contents of the egg do not completely fill the eggshell in this region. This cavity is called the *air space*. Describe the position of this air space.

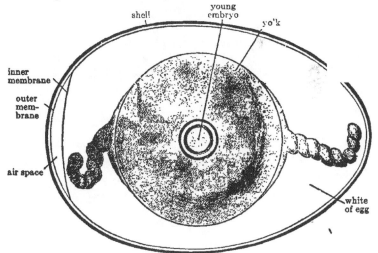

FIG. 52. — Egg of a hen.

70   ANIMAL BIOLOGY

   3. Pick off the pieces of shell and allow all the contents of the shell
       to flow out into a cup. or deep saucer, taking care not to
       break the yolk.
      a. State what has been done, and describe the position and color
          of the white and of the yolk of the egg.
      b. Note two twisted strands extending from the yolk towards
          each end of the egg. These help to protect the yolk from
          sudden jars. Describe the position, appearance, and use
          of these strands.
   4. Carefully turn the yolk until you notice a white spot. This spot
       is the beginning of an *embryo chick*.
      Describe the position and appearance of a young chick embryo.

54. Reproduction and life history. — In the preceding section we have seen that a bird's egg consists of a hard shell, a membrane, the white and the yolk; and that on the outer surface of the latter is a tiny *embryo*. Let us now see how this egg is formed and developed.

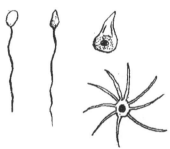

FIG. 53. — Sperm-cells of various animals.

In our study of seed-plants we learned that plant embryos are formed in the ovary of a pistil after an egg-cell has been fertilized by a sperm-cell. In the case of insects (**22, 27**) and the fish (**104**) we find that egg-cells are produced in organs of the female known as *ovaries* and that before an egg-cell can develop into an embryo (except in rare cases) it must be fertilized by a sperm-cell (Fig. 53) which has been formed in the *spermary* of a male.

   If the ovaries of a hen are examined, they will be found to consist of a large number of spherical objects, the larger ones being yellow, which vary in size from tiny dots to full-sized yolks (Fig. 54). If any one of these is examined carefully with a microscope, a single egg-cell may be found. After the yolk has attained its full size and the egg-cell has been

Fig. 54. — Ovary of hen, and egg in egg-tube.

fertilized, it receives its coating of white, and the whole is covered with the membranes and the shell.

Immediately after fertilization takes place, by the process of cell division many cells are formed. At the time the egg is laid, the chick embryo appears as a tiny white spot on the surface of the yolk when the egg is opened (**53**, 4). Further development of this embryo, however, cannot take place unless the egg is kept warm. This is brought about when

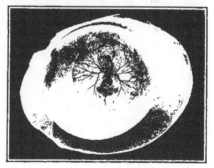

Fig. 55. — Egg of hen, showing embryo chick on surface of yolk. (Beebe, "The Bird.")

the hen broods over the eggs. Gradually the cells of the different external and internal organs are formed (Fig. 55) from the food material furnished by the yolk and the white of the egg, and at the end of three weeks the young chick breaks through the shell, and soon, under the protection of the mother hen, begins to search for food. When first hatched, the feathers are relatively small and downy. The further development of the chick is largely a matter of growth in size and of change in the character of the feathers.

55. Nests and care of young. — The method of reproduction in all birds is much the same as that already described

Fig. 56. — Comparative size of the eggs of ostrich, hen, and humming bird. (Photographed by E. R. Sanborn, N. Y. Zoölogical Park.)

for the chick. Many birds, however, are much more help less when they emerge from the egg than are chickens, and so they are sheltered in *nests*, and the food of the young birds is brought to them by their parents until they are able to fly and for several days afterwards.

BIRDS 73

Nests differ greatly in their complexity and in the kind of material used. Some birds, for example the gulls and many other sea-birds, usually deposit their eggs on rocky ledges or in slight depressions in the sand along the shore. On the other hand, the Baltimore oriole constructs out of grasses, plant fibers, and strings a marvelous nest hanging high up in the trees, near the outer ends of branches (Fig. 73). Between these two extremes are all gradations of nest complexity.

The eggs laid by birds vary in number, size, and color. The tiny humming bird, for instance, lays two white eggs, each a third of an inch in diameter (Fig. 56); three to five greenish blue eggs, each nearly an inch in diameter, are usually found in a robin's nest, while an ostrich deposits twelve to fourteen eggs, each weighing three to four pounds.

56. Common methods of classification. — One of the simplest ways of classifying birds is that of dividing them into groups according to the *kind of food* they eat. For instance, we may speak of fish-eating, seed-eating, and insect-eating birds. This, however, is far from being a scientific classification, since birds that differ considerably in struc-

FIG. 57. — The pelican. (Photographed by E. R. Sanborn.)

ture, and therefore not closely related, frequently live upon the same kind of food. For example, both the pelican (Fig. 57)

Fig. 58. — Belted kingfisher. (Wright's 'Citizen Bird.")

Fig. 59. — Herring gull. (Wright's "Citizen Bird.")

and kingfisher (Fig. 58) catch and eat fish for food, yet a glance at the two figures shows how unlike in form these two birds are.

A second scheme of classification is that based upon their *habitat*. Thus we may speak of water birds, shore birds, marsh birds, and land birds. This plan, too, may group together birds strikingly unrelated in structure and habits, as becomes clear when we compare two land birds like the hawk (Fig. 64), and the sparrow (Fig. 70).

57. Scientific classification of birds. — Modern scientific classification divides the birds of North America into seventeen groups or

FIG. 60. — Blue heron. (Wright's "Citizen Bird.")

orders, all the birds of a given order resembling each other more or less in structure. The common names given to some of these orders are suggested by their habits. As examples we may name *diving birds* (loon), *long-winged swimmers* (gulls and terns), *scratching birds* (hens, turkeys, and quails), *birds of prey* (eagles, hawks, and owls), and *woodpeckers* (downy woodpecker). The highest order, known as the *perching birds*, is divided into twenty *families*, some of which are the *crow family*, the *sparrow family*, the *warbl*er, and the *thrush family.* The total number of species of the perching birds is far greater than that of all other species taken together. We shall now group together a few of the more closely related orders, and discuss somewhat their characteristic adaptations of structure.

Fig. 61.— Flamingoes. (Photographed in N. Y. Zoölogical Park, by E. R. Sanborn.)

58. Webfooted birds (swimming birds). — In this group we include several orders of birds that have webbed feet, which fit them for swimming in the water. Common examples of such birds are ducks, geese, albatross, and gulls (Fig. 59). Near the tail region of most of these birds an oil gland is developed, from which the bird obtains the oil that it uses in keeping its feathers from getting water-soaked; this is likewise true of all other birds. As one would expect, a large number of these species feed upon fish and other water animals.

59. Wading birds. — All the birds in this group have long, slender legs, which adapt them for wading out into the water for food. Such birds are the herons (Fig. 60), egrets, storks, and cranes. The flamingoes (Fig. 61) have webbed feet like swimming birds, and so they are regarded as connecting links between swimming and wading birds.

Fig. 62. — Bobwhite.

60. Scratching birds. — This group includes the domesticated chicken and turkey and the quail (Fig. 62). All our various forms of chickens are descended from the

Fig. 63. — Male and female jungle fowl. (Photographed by E. R. Sanborn, from specimens of the American Museum of Natural History.)

small jungle fowl of India (Fig. 63). The wild turkey still exists in some parts of our country, but it is being rapidly exterminated by hunters. The toes of all the scratching birds are armed with strong, blunt nails, by which they are enabled to dig in the soil for insects and worms. All these birds, too, feed to some extent upon grain.

Fig. 64. — Red-shouldered hawk.

61. Birds of prey. — The hawks (Fig. 64), eagles, and owls (Fig. 65), which comprise this group have acquired the name of birds of prey from their habit of catching and feeding on rats, mice, birds, and other animals. Their feet are armed with sharp incurved claws, and the upper part of their bills is hooked; and so they are specially adapted for seizing and tearing their prey.

62. Woodpeckers. — These birds are admirably adapted to creep and climb up the trunks of trees, for they have two clawed toes extending forward, and two backward, and their tail feathers are so stiffened that they serve as props against the bark when the bird is resting (Fig. 66). The food of the woodpeckers is largely composed of insects, which these birds secure by digging them out of the bark or the wood with their stout, chisel-like bills, and then spearing them with their long tongues.

Fig. 65. — Short-eared owl. (Wright.)

Fig. 66. — Downy woodpecker. (Wright.)

63. Perching birds. — This order, as we have said before, contains by far the largest number of species of birds. All these birds are specially adapted for holding to the limbs of trees, since the mechanism of the leg is so arranged that the toes are automatically clutched to the support upon which the bird is sitting. In this group are included practically all of our bird vocalists, hence the perching birds are often called the " song birds." Among the most beautiful of our songsters are the bobolinks (Fig. 67), catbird, and thrushes (Fig. 68).

The young of all the perching birds, for weeks after they are

FIG. 67. — Male and female bobolink.

hatched, are helpless in the nests and are unable to feed themselves. Most of the food of young birds consists of the larvæ of insects and some of the families, *e.g.* the fly-catchers (Fig. 69), feed upon insect food throughout their life. The sparrow family (Fig. 70), on the other hand, choose largely a diet of seeds. Almost every kind of food, however, is eaten by some of the perching birds.

64. Migration of birds. — Some of the birds like the chickadee and downy woodpecker, remain in the middle and northern United States throughout the year, and hence are known as *permanent residents* of these regions. Many birds, however, spend the winter

FIG. 68. — Wood thrush.

in the warmer regions of the South and in the spring months move northward; some of them, like the robin (Fig. 71) and the bluebird, build their nests, rear their young, and stay all summer in northern and middle United States. Such birds are called *summer residents*. Still other birds rear their young in Canada and even farther north, and come to us only as *winter visitants*. This seasonal movement of birds is known as *migration*. Migration is

FIG. 69. — Kingbird. (Courtesy of National Audubon Society.)

FIG. 70. — Tree sparrow. (Courtesy of National Audubon Society.)

especially characteristic of the perching birds. For this reason, the birds in this, the highest order, are known as "birds of passage."

FIG. 71. — The robin.

Date When Seen	Place Where Seen	Size[1] Compared to Robin or Sparrow	Color of Back	Color of Breast	Other Striking Colors	Name of Bird
Mar. 10, 1912	Lower limbs of trees	Larger than sparrow	Bright blue	Reddish brown	Belly white	Bluebird

65. Field work on birds. — Pupils should become familiar with the size, form, colors, and song of as many birds as possible, and should note carefully where each kind of bird is most commonly found (*e.g.* in marshes, trees, bushes, or on the ground). In this study bird glasses or opera glasses are very useful. Books like

---

[1] Length of robin from tip of bill to tip of tail feathers, about 10 inches; length of sparrow from tip of bill to tip of tail feathers, about 6 inches.

Chapman's "Bird Life," Wright's "Citizen Bird" and "Birdcraft," Hornaday's "American Natural History," should be frequently consulted. In order to record striking characteristics as a help toward identifying birds, it is suggested that each pupil fill out a table as shown on page 82.

66. Importance of birds to man. — Few animals are more beautiful in form and color than are many of our most common birds, and one of the greatest delights of springtime is to greet the return of the bluebirds, tanagers, thrushes, and others of our feathered friends. " To appreciate the beauty of form and plumage of birds, their grace of motion and musical powers, we must know them. Once aware of their existence, and we shall see a bird in every bush and find the heavens their pathway. One moment we may admire the beauty of their plumage, the next marvel at the ease and grace with which they dash by us or circle high overhead The comings and goings of our migratory birds in springtime and fall, their nest-building and rearing of young, their many regular and beautiful ways as exhibited in their daily lives, stir within us impulses for kindness toward the various creatures which share the world with us. . . . But birds will appeal to us most strongly through their song. When your ears are attuned to the music of birds, your world will be transformed. Birds' songs are the most eloquent of Nature's voices: the gay carol of the grosbeak in the morning, the dreamy, midday call of the pewee, the vesper hymn of the thrush, the clanging of geese in springtime, the farewell of the bluebird in the fall, — how clearly each one expresses the sentiment of the hour or season!" — Quoted from Bulletin No. 3 of University of Nebraska, and from Chapman's "Bird Life."

The value of birds to man as objects of beauty cannot be measured, it is true, in dollars and cents; but were we to lose the birds, we should realize all too well how much they contribute to the happiness of every lover of nature. When, however, we come to discuss the economic value of birds, the good that they do cannot be overestimated. Biologists have carried on long series of studies to determine accurately the food of different kinds of birds. This has been done by watching them while they are eating or while feeding their young, and by examining the contents of birds' stomachs. The following paragraphs contain descriptions of some of the ways in which birds are of inestimable use to man.

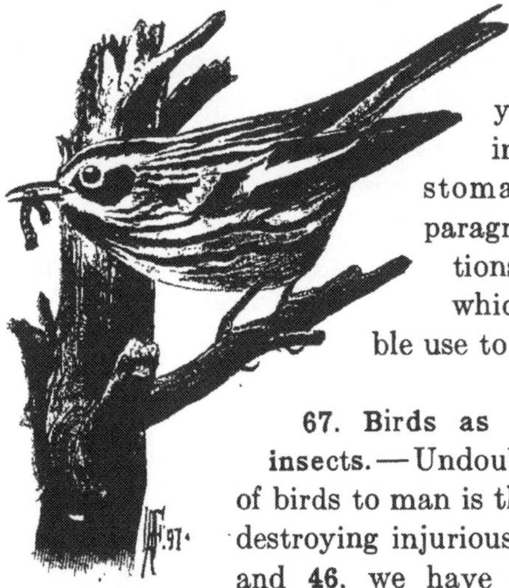

FIG. 72. — Black and white warbler.

67. **Birds as destroyers of harmful insects.** — Undoubtedly the greatest value of birds to man is the good that they do in destroying injurious insects. In **13–18, 23,** and **46**, we have described some of the ravages made by our insect foes.

"But if insects are the natural enemies of vegetation, birds are the natural enemies of insects. In the air swallows and swifts are coursing rapidly to and fro, ever in pursuit of the insects which constitute their sole food. When they retire, the nighthawks and whip-poor-wills will take up the chase, catching moths and other nocturnal insects which would escape day-flying birds. Fly-catchers (Fig. 69) lie in wait, darting from ambush at passing prey, and with a suggestive click of the bill returning to their post. The warblers (Fig. 72), light, active creatures, flutter about the terminal foliage, and with almost the skill of a humming bird, pick insects from the leaf or blossom. The vireos patiently explore the underside of leaves and odd nooks and corners to see that no skulker escapes. The woodpeckers (Fig. 66), nuthatches, and creepers attend to the trunks and limbs, examining carefully each inch of bark for insects' eggs, and larvæ, or excavating for the ants and borers they hear within. On the ground the hunt is continued by the thrushes (Fig. 68), sparrows (Fig. 70), and other birds that feed upon the innumerable forms of terrestrial insects. Few places in which insects exist are neglected; even some species which pass their earlier stages or entire lives in the water are preyed upon by aquatic birds." [1] — From CHAPMAN'S " Bird Life."

As examples of the number of insects destroyed by individual birds we may give the following: Six robins in Nebraska ate 265 Rocky Mountain locusts; the stomachs of four chickadees contained 1028 eggs of cankerworms; 101 potato beetles were found in the stomach of a single quail (Fig. 62); and 250 hairy caterpillars, which other birds do not eat, were devoured by a yellow-billed cuckoo. (Frontispiece.)

68. Birds as destroyers of weed seeds. — Another way in which birds are useful to man is in the destruction of weed seeds. Most perching birds that feed largely upon seeds, *e.g.* the sparrows and finches, have stout, conical bills (Fig. 70) which are specially adapted for crushing seeds. In one of the pamphlets of the United States Department of Agriculture, entitled " Some Common Birds and their Relation

[1] Before assigning this section for study each of the birds named should if possible be shown to the class, or at least colored pictures of the birds, *e.g.* in Chapman's "Bird Life."

Fig. 73. — Nest of Baltimore oriole; male bird below, female above. (Photographed by A. E. Rueff of the Brooklyn Institute of Arts & Sciences.)

to Agriculture,". the writer estimated that in the state of Iowa during the six months of fall and winter, tree sparrows devoured 875 *tons of weed seed*. An actual count of the stomach contents of a bobwhite showed the presence of 400 pigweed seeds. In the stomach of another were 500 seeds of ragweed.

69. Birds as destroyers of rats and mice. — We learned in 61 that hawks and owls by their hooked bills and claws are admirably fitted to clutch and tear living prey. It has been demonstrated that the food of many of these birds consists almost wholly of small gnawing mammals (*e.g.* field mice) (Fig. 64) which are exceedingly injurious in fields of grain. An examination of the stomachs of fifty short-eared owls (Fig. 65) showed that 90 per cent of them contained nothing but mice. Forty of the forty-nine stomachs of the rough-legged hawks were found to contain mice, while most of the rest contained injurious animals.

70. Birds as scavengers. — Some birds of prey, like the turkey buzzards of the Southern states, eat animals that are dead. " These animals may be seen at all hours of the day sailing through the air in majestic circles or lazily resting on stumps or trees after a feast of their filthy food. They perform an important service as scavengers, disposing of all sorts of animal matter that would pollute the air. On this account they are seldom molested by man and in some States are protected by law. They devour both fresh and putrid meat. They are known sometimes to capture live snakes and to attack helpless animals of many kinds. Along the seashore they feed upon dead fish cast up by the waves." — WEED and DEARBORN, " Birds in their Relations to Man." Gulls (Fig. 59) also serve a useful purpose by devouring dead fish and other refuse along our coast line and in our harbors.

71. Birds injurious to man. — We have discussed briefly in the preceding sections some of the ways in which birds are of incalculable value to man. It must be admitted, however, that some birds are of doubtful value, while others are positively injurious. As an example of a bird, which, to say the least, is a nuisance, we may mention the common English sparrow. This bird was first introduced from England into the United States in Brooklyn, N. Y., in 1851, because it was expected to attack some of our injurious insects. These sparrows have multiplied so rapidly that now they are found practically everywhere in the United States. " As destroyers of noxious insects, the sparrows are worse than useless." Thus, for instance, the stomach of a single cuckoo (Frontispiece) was found to contain more insects than did the stomachs of 522 English sparrows.

But even more serious are the positive charges that have been proved against this bird. It pecks at and destroys the young buds of trees, and later injures many fruits while they are ripening. It causes great losses in the grain fields from the time of planting to that of harvesting; and worst of all is the fact that it molests and drives away our native song and insect-eating birds.

The crow (Fig. 74) is another bird that on the whole is probably more injurious than beneficial for the following reasons: " (1) Crows seriously damage the corn crop, and injure other grain crops, usually to a less extent. (2) They damage other farm crops to some extent, frequently doing much mischief. (3) They are very destructive to the eggs and young of domesticated fowls. (4) They do incalculable damage to the eggs and young of native birds. (5) They do much harm by the distribution of seeds of poison ivy, poison sumach, and perhaps other noxious plants. (6) They do much harm by the destruction of beneficial insects. On the other hand: (1) They do much good by the destruction of injurious insects. (2) They are largely beneficial through their destruction of mice and other rodents. (3) They are valuable occasionally as scavengers." — W. B. BARROWS, " The Food of Crows."

FIG. 74. — The crow.

While most of the hawks are undoubtedly beneficial (69), two species, namely, Cooper's hawk and the sharp-shinned hawk, must be kept down to limited numbers. Both of these are " chicken-hawks," and in addition they ruthlessly destroy great numbers of our most valuable wild birds.

72. Summary of the relation of birds to human welfare. — Library study.

For further facts like the following, consult, Weed and Dearborn's "Birds and their Relation to Man," Forbush's "Useful Birds and their Protection," Hornaday's "American Natural History," pamphlets of Department of Agriculture (which may be obtained free from Washington, D.C., and from State Departments of Agriculture), and articles on birds and insects in Encyclopedias.

Name of Bird	Time of Visitation	Kind of Animal Food Eaten	Kind of Vegetable Food Eaten	Remarks
Robin	Summer resid.	Insects, 42 per cent	Small fruits and berries, mostly wild, 58 per cent	Beneficial
Phœbe	Summer resid.	Insects caught on wing, 93 per cent	Wild fruits, 7 per cent	Beneficial
Hairy woodpecker	Perm. resid.	Wood-boring insects and ants	Wild fruits	Beneficial
Yellow-billed cuckoo	Summer resid.	Insects, largely hairy caterpillars		Beneficial
Quail or bobwhite	Perm. resid.	Insects in summer	Weed seeds during rest of year	Beneficial
Tree sparrow	Winter visit.		Weed seeds	Beneficial
Bobolink	Summer resid.	Insects in North	Rice in South	$2,000,000 loss to rice crop
Short-eared owl	Perm. resid.	Rats, mice and other small mammals		Beneficial

Name of Bird	Time of Visitation	Kind of Animal Food Eaten	Kind of Vegetable Food Eaten	Remarks
Sparrow-hawk	Summer resid.	Mice and insects		Beneficial
Cooper's hawk	Summer resid.	Poultry and song birds		Injurious
Sharp-shinned hawk	Summer resid.	Poultry and song birds		Injurious
Crow	Perm. resid.	Insects, mice, eggs and young of other birds	Corn and other crops, weed seeds	Doubtful value
English sparrow	Perm. resid.	Insects rarely	Buds, fruit, grain	Drives away useful birds

73. Causes of decrease in bird life. — Certainly enough has been said to show that when all things are considered birds are exceedingly useful to man. One would therefore expect that every possible means would be taken to protect all kinds of valuable birds. Yet what do we find? "To-day the first thing to be taught is the fact that from this time henceforth *all birds must be protected, or they will all be exterminated.* To-day, it is a safe estimate that there is a loaded cartridge for every living bird. Each succeeding year produces a new crop of gun-demons, eager to slay, ambitious to make records as sportsmen or collectors. If a bird is so unfortunate as to possess plumes, or flesh which can be sold for ten cents, the mob of pot-hunters seeks it out, even unto the ends of the earth." — Hornaday's "The American Natural History."

A careful investigation made in 1897 for the New York Zoölogical Society showed that during the fifteen years between 1883 and 1898 in all but four states [1] the number of birds had strikingly decreased. For example, in New York State the decrease was 48 per cent, or almost one half; in Florida it was over three fourths, while the average for the whole country was 46 per cent. Among the principal reasons given by the 180 careful observers who assisted Dr. Hornaday in the foregoing inquiry were the following : " (1) sportsmen and so-called sportsmen, (2) boys who shoot, (3) market hunters and pot-hunters, (4) plume-hunters and milliners' hunters, . . . (6) egg-collecting, chiefly by small boys, (7) English sparrow, . . . (9) Italians, and others, who devour song birds."

74. Destruction of birds by cats. — " As the cat is not an actual necessity, and as it is a potent carrier of contagious diseases, which it spreads, particularly among children, it would be far better for the community if most of the bird killing cats now roaming at large could be painlessly disposed of.   . Where the cat is deemed necessary in farm or village, no family should keep more than one good mouser, which should never be allowed to have its liberty during the breeding season of birds.   Cats can be confined during the day in outdoor cages as readily as rabbits, and given the run of the house at night." — FORBUSH, " Useful Birds and their Protection."

75. Destruction of birds by boys. — One of the most serious menaces to our native bird life is the small boy who has the " egg-collecting fever." All the eggs he can find in his keen-eyed searches through the woods and fields are

[1] Kansas, Wyoming, Utah, and Washington were the only states that showed an increase in bird life.

destroyed to increase his collection. If this served any really useful purpose, the resulting wholesale destruction of birds might possibly have some justification. But ninety-nine out of a hundred of these collections are soon forgotten and become useless without having made any real contribution to the knowledge of the possessor.

The small boy, too, unfortunately carries his destructive work among birds still further, as the following typical incident will show. A biologist reports meeting near Washington, D.C., "one such youngster, and upon examining his game bag found it absolutely full of dead bodies of birds which he had killed since starting out in the morning. One item alone consisted of seventy-two ruby and golden-crowned kinglets. The fellow boasted of having slain over one hundred catbirds that season."

76. Destruction of birds for food. — In the early days of the white settlements in North America, the game birds like the grouse and duck were abundant and they were of necessity killed, as were other wild animals, for food. Later on began the killing of birds for sport. As the forests were cut down, the birds had less and less protection, and had not legislation intervened, the game birds would long since have been exterminated. As it is, they have been killed faster than they breed; and this means ultimate extermination.

To this destruction of game birds for food, in more recent times has been added the wholesale slaughter of many of our smaller birds like the thrushes, sparrows, warblers, and woodpeckers. It is claimed that this has been largely due to the demands of our immigrant population in the North and to the negroes in the South. "However, there is scarcely a hotel in New Orleans," says Professor Nehrling, "where small birds do not form an item on the bill of fare. At certain seasons the robin, wood thrush, thrasher, olive-backed thrush, hermit thrush, chewink, flicker, and many of our beautiful sparrows form the bulk of the victims; but catbirds, cardinals, and almost all small birds, *even swallows*, can be found in the markets."

77. Destruction of birds for millinery purposes. — Even more ruthless than the slaughter of birds for food by boys and by men is that caused by the demand for birds for millinery purposes. Here the final responsibility rests upon women alone. A single dealer in the South declared that in the course of a single year he handled 30,000 bird skins, the largest part of which were used in the decoration of hats.

FIG. 75. — Egret, nest, and young. (Courtesy National Audubon Society.)

The Florida egret heron (Fig. 75) has been practically exterminated for this purpose "Twenty years ago," says Chapman, "it was abundant in the South, now it is the rarest of its family. The delicate 'aigrettes' which it donned as a nuptial dress were its death warrant. Woman demanded from the bird its wedding plumes, and man has supplied the demand. The Florida herons or egrets have gone, and now he is pursuing the helpless birds to the uttermost parts of the earth. Mercilessly they are shot down at their roosts or nesting grounds, the coveted feathers are stripped from their backs, the carcasses are left to rot, while the young in the nest above are starving."

" This slaughter of the innocents is by no means confined to the Southern states. During four months 70,000 bird skins were supplied to the New York trade by one Long Island village. 'On the coast line of Long Island,' wrote

FIG. 76. — Tern.

Mr. William Dutcher, not long ago, 'the slaughter has been carried on to such a degree that, where, a few years since, thousands and thousands of terns (Fig. 76) were gracefully sailing over the surf-beaten shore and the wind-rippled bays, now one is rarely to be seen.' Land birds of all sorts have also suffered in a similar way, both on Long Island and in adjacent localities in New Jersey. Nor have the interior regions of the United States escaped the visits of the milliner's agent. An Indianapolis taxidermist is on record with the statement that in 1895 there were shipped from that city 5000 bird skins collected in the Ohio Valley. He adds that 'no county in the state is free from the ornithological murderer,' and prophesies that birds will soon become very scarce in the state.

"These isolated examples can only suggest the enormous number of birds that are sacrificed on the altar of fashion. The universal use of birds for millinery purposes bears sufficient testimony to the fact. Yet it is probable that most women who follow the fashion seldom appreciate the suffering and the economic losses that it involves." — WEED and DEARBORN, "Birds in their Relations to Man."

78. Effects of bird destruction — While the æsthetic loss to mankind resulting from the destruction of our wild birds cannot, as we have said, be computed, yet even in the cities this loss is beginning to be realized as we see the song birds in the parks steadily diminishing in number. Everyone, however, is affected by the increasing cost of our food supply, and we have but to review the facts stated in the preceding sections to show that the destruction of our wild birds has a very important bearing on the present situation.

Every farmer knows that it is impossible to raise the crops of a single year without battling with insect pests. The time and expense involved in applying insect-destroying preparations would be difficult to compute, and even after the year's contest is ended, the insects are often victorious. In ruthlessly destroying the wild birds man has interfered with the "balance of nature" and so has helped the ravaging hordes of insects and gnawing animals to multiply without adequate check. All this means that we, the consumers of the fruits, the vegetables, and the grains, must pay higher prices for the food we eat and the clothes we wear.

79. Conservation of birds. — But it is not yet too late to save the remnant of the birds still left to us, and even to increase the bird life of our country. It is evidently necessary, however, in the first place, that laws similar to the following should be passed in every state.

"The Bird Law of the American Ornithologists' Union. — An Act for the Protection of Birds and their Nests and Eggs.

"Section 1. — No person shall within the State of —— kill or catch or have in his or her possession, living or dead, any wild bird other than a game bird, nor shall purchase, offer, or expose for sale any such wild bird after it has been killed or caught. No part of the plumage, skin, or body of any bird protected by this section shall be sold or had in possession for sale.

"Section 2. — No person shall within the State of —— take or needlessly destroy the nest or the eggs of any wild bird nor shall have such nest or the eggs in his or her possession.

"Section 3. — Any person who violates any of the provisions of this act shall be guilty of a misdemeanor, and shall be liable to a fine of five dollars for each offense, and an additional fine of five dollars for each bird, living or dead, or part of bird, or nest and eggs possessed in violation of this act, or to imprisonment for ten days, or both, at the discretion of the court.

"Section 4. — Sections 1, 2, and 3 of this act shall not apply to any person holding a certificate giving the right to take birds and their nests and eggs for scientific purposes, as provided for in Section 5 of this Act. . .

"Section 7. — The English or European house sparrow (Passer domesticus) is not included among the birds protected by this act."

"In addition to the above, every state that has not already done so, should at once enact laws to prohibit the sale of all wild game at all seasons, and to stop all shooting of game in late winter and spring. About one half the states have done this, and the other half should act without delay. The sale of game has almost destroyed our once magnificent supply of game birds. We have no right to hand down to posterity a gameless continent. The wild life of to-day is not wholly ours to dispose of as we please. It has been given to us *in trust*. We must account for it to those who come after us and audit our records."[1] — Dr. W. T. HORNADAY.

But laws, however stringent, are of little avail unless there is a healthy public sentiment to bring about their enforcement. Thus, for instance, it is evident that laws merely designed to prevent the killing of birds for millinery purposes will be ineffective, so long as women are permitted to wear birds. One thing will completely stop the cruelty of bird millinery — the disapprobation of fashion. " It is our women who hold the great power. Let our women say the word, and hundreds of bird lives will be preserved every year. And until woman does use her influence it is vain to hope that this nameless sacrifice will cease until it has worked out its own end and the birds are gone." — WEED and DEARBORN, " Birds in their Relations to Man."

80. What boys and girls can do to protect birds. — " Now that adequate statutes are either enacted or may reasonably be expected very soon, it remains to scatter information about birds everywhere, so that laws may be respected and it is in this line that those interested in their conservation should work. There must be lectures, short articles of a popular nature in newspapers and magazines, distribution of government and other publications relating to birds,

---

[1] The authors are indebted to Dr. W. T. Hornaday, of the New York Zoölogical Park, for many suggestions relating to conservation of birds and for a careful reading of the chapters on birds and fishes.

posting bird laws in conspicuous places, and most important of all, systematic bird work in public schools. The importance of engaging the interest of our youth in birds cannot be overestimated. It results in a double benefit, for the birds will be held in higher esteem and the children will become possessed of a source of lasting pleasure. The nest-robbing, bird-shooting boy and the feather-wearing girl may be made the friends and allies of the birds." — WEED and DEARBORN, "Birds in their Relations to Man."

But not only should the boy cease to destroy nests and shoot birds; not only should the girl cease to wear any part of a wild bird; but boys and girls alike should do all they can to induce others to do likewise. Much may also be done, likewise, even in the vicinity of large towns, to attract birds and induce them to nest. In the first place, the nests and eggs of the English sparrow should be destroyed whenever found. Stray cats should be kept from harming birds. Pieces of meat, bones, and suet, when hung in the trees in winter time, and crumbs and grains scattered about, will serve to attract the winter visitants and, when thus attracted, these birds devour great numbers of the eggs and insects in the hibernating stages that during the following season would attack the fruit and shade trees. And finally, any ingenious boy can construct and put in the trees bird houses that in the springtime would become the

FIG. 77. — Bird house made by a twelve-year-old boy.

nesting places of bluebirds, wrens, tree swallows, and martins.[1] (Fig. 77.)

[1] For other methods of encouraging birds see Weed and Dearborn, "Birds in their Relations to Man," pp. 304–315. Trafton, "Methods of Attracting Birds," and leaflets of National Association of Audubon Societies, 1974 Broadway, N. Y., *e.g.* No. 16 (Winter Feeding of Wild Birds) and No. 18 (Putting up Bird Boxes), 1 cent each.

CHAPTER III

FROGS AND THEIR RELATIVES

81. Study of the frog. — Laboratory study.

A. *Regions and appendages.* — The frog's body consists of two principal parts, or regions; namely, the *head* and *trunk*. The line of union between the two regions is just in front of the anterior appendages (arms).
  1. Locate the appendages (*arms* and *legs*) attached to the trunk.
  2. Name and locate the organs that you find on the head, giving the number of each.

B. *Breathing organs.*
  1. Describe the location of the nostrils on the head.
  2. Examine a preserved specimen in which a stiff bristle has been passed through one of the nostrils.
     a. Tell what was done.
     b. Into what cavity has the bristle emerged?
     c. (Optional.) What is one difference, therefore, between the nostrils of a fish and of a frog?
     d. In what region (anterior or posterior) of the roof of the mouth cavity are the inner openings of the nostrils located?
  3. (Demonstration.) Just back of the tongue there is a narrow opening that leads into the windpipe (*trachea*). This opening is called the *glottis*.
     a. Locate the glottis.
     b. Does the glottis extend lengthwise or crosswise of the mouth cavity?
     c. Into what does the glottis open?
  4. Examine a dissected frog prepared in such a way as to show the lungs.

>
> a. State the location of the lungs with reference to the head and the cavity of the trunk.
> b. Describe the appearance of the lungs.
> c. Insert the end of a glass tube, that has been drawn out to a small diameter, into the glottis opening and blow air into the lungs. Describe what you have done and state the result.
>
> 5. Name in order the openings, the cavity, and the tube through which the air must pass in order to reach the lungs.

C. *Breathing movements.*
> 1. Place a frog in a glass jar with an inch or two of water and watch the action of the floor of the mouth. This is one of the breathing movements. Describe this breathing movement of the frog.
> 2. There are two breathing movements of the sides of the trunk, one a very active inward and outward movement, and the other a very slight inward and outward movement. When you have seen these two kinds of movements of the sides, describe them and state which kind occurs the more frequently.

D. *How the frog exhales.*
> 1. What effect will the active inward movement of the sides have upon —
>   a. The size of the body cavity?
>   b. The size of the lungs?
>   c. The pressure of the air in the lungs?
> 2. When the sides of the trunk move actively inward, will the air move into the lungs or out? Why?
> 3. Through what passages will the air go from the lungs to the outside of the frog?

E. *How the frog inhales.*
> 1. When the floor of the mouth moves downward —
>   a. Will the size of the mouth cavity be made larger or smaller?

FROGS AND THEIR RELATIVES

    b. If the nostrils are now open will the air move into the mouth cavity or out? Why?
  2. When the floor of the mouth is raised —
    a. Will the size of the mouth cavity be increased or decreased?
    b. Will the pressure of the air in the mouth cavity be increased or decreased?
    c. If both the nostrils and the glottis are now open, in what directions will air be forced?
    d. What causes the slight outward movements of the sides of the trunk in the region of the lungs?

F *How the lungs are fitted for breathing organs.* (Suggested as home work.)

    When the lungs are inflated (see *B*, 4 above) they look like bags (Fig. 80). The lungs are hollow, and their walls are composed of thin material. In these membranous walls are thin-walled blood vessels known as *capillaries*. The heart forces blood that has come from the body into these capillaries of the lungs, and then back to the heart. Bearing in mind that respiration in animals is essentially the same as in plants (**P. B.**, 82) —
  1. State what waste substance the blood brings to the lungs to be given off from the capillaries.
  2. What gas will the blood in the capillaries take up from the air in the lungs?
  3. How are the walls of the lungs and of the capillaries of the lungs fitted by structure to make this interchange of gases possible?

G. *Food-getting.*

    To the Teacher.—Select a number of as large preserved or freshly killed frogs as you can get. Open the jaws as far as possible and keep them in this position by means of small pieces of wood.

1. Seize the posterior or hind end of the tongue and pull it forward.
    a. Tell what you have done and state which end of the tongue is attached to the floor of the mouth.
    b. Describe the shape of the tip end of the tongue.
2. The living frog can extend its tongue much farther than you have been able to do in the case of the preserved frog, and in the living frog the tongue is covered with a very sticky substance. The tongue is used to catch insects at some distance from the animal (Fig. 79). Tell how you think the frog could use its tongue to catch insects and get them into its mouth.[1]
3. Look for teeth on the jaws of a skeleton of a frog, or if you cannot obtain a skeleton, rub the finger over the jaws of a preserved specimen.
    a. Which of the jaws has teeth?
    b. Describe the location of the teeth on the jaw.
    c. State the shape and size of the *jaw teeth*.
    d. What is the probable use of the jaw teeth?
4. (Optional.) Look on the roof of the mouth for two relatively large *palate teeth*. Rub the fingers over the surface of the palate teeth.
    a. Tell what you have done.
    b. What have you found out about the palate teeth?
    c. What is the probable use of the palate teeth?

H. *How a frog swallows.*

1. Gently touch the eyes of a living frog until it draws them into the head. Tell what you have done and observed.
2. Look at the roof of the mouth of a preserved specimen while you push the eyes into the head. Tell what you have done and describe the effect produced in the roof of the mouth.
3. How will the act of pushing the eyes into the head be useful to a frog in swallowing?[1]

[1] If possible live frogs should be fed on meal worms, or other insects, and the feeding movements observed.

I. *(Optional.) Sketch of the mouth cavity.*
   1. Open wide the mouth of a large frog and make a sketch to show the shape of the mouth cavity twice the natural size, and the shape and thickness of the upper and lower jaws.
   2. Draw the following parts to show their location, size, and shape: jaw teeth, palate teeth, tongue, glottis, nostril openings, swellings caused by the eyes.
   3. Farthest back in the throat find an opening that extends crosswise. It is the opening into the *gullet* and is just behind the glottis. Push the handle of the forceps into this opening and draw it (in your sketch) partly opened.
   4. Label upper jaw, lower jaw, jaw teeth, palate teeth, tongue, glottis, opening of gullet, nostril openings, swellings caused by eyes.

J *Structure of arms and leg*

   Place a frog in a glass jar at least half full of water to cause the animal to extend the hind legs.
   1. Make a sketch (natural size) of an arm to show the shape and size of the following parts: upper arm, elbow, forearm, hand, number of fingers. Label each part.
   2. Draw one of the legs (natural size) to show the following parts: thigh (next the body), knee, shank, ankle (elongated region above foot), foot, toes, web between toes. Label each part.

K. *How a frog swims.*

   Place an active frog in a sink or other receptacle large enough to afford it room to swim. The water should be deep enough so that the frog will not strike the bottom with the legs. Get the frog to swim the full length of

the receptacle as many times as may be necessary to answer the following: —
1. Tell what you have done.
2. Describe the movements of the hind legs in swimming.
3. In which of these movements are the toes spread out?
4. In which of these movements, therefore, can the frog get the best hold upon the water?
5. In which direction must the frog push the harder in order to move in the direction that it does?
6. In what respects are the posterior appendages well fitted for swimming?
7. In what respects are the anterior appendages not as well fitted as the legs for swimming?

L. *How a frog jumps.*

Place a frog where there is plenty of room, and get it to jump as many times as necessary to answer the following: —
1. Tell what you have done.
2. Describe the position of the parts of the legs just before the frog jumps.
3. Describe the two movements made by the parts of the leg in the act of jumping and when about to land.
4. In which of these two movements must the frog use the greater force?
5. Which movement, therefore, throws the frog into the air?
6. In what respects are the legs better adapted for jumping than the arms?

N *Internal organs.*

To the Teacher. — Put into a covered jar enough frogs to supply each two students with a specimen. Put into the jar some ether, or better, saturate a small sponge with the ether and place it in the jar. When the animals are dead, dissect them as follows: lift up the skin of the ventral wall of the abdomen with

FROGS AND THEIR RELATIVES

the forceps; carefully insert the point of the scissors near the posterior end of the trunk, and carefully cut forward on one side of the body as far as the tip of the head, and back on the other side of the trunk, until the skin is completely removed from the ventral surface. In a similar manner remove the muscular wall that covers the trunk, being careful not to injure the internal organs. If time allows, remove also the skin from one leg; call attention to the thinness of the skin and to the underlying blood vessels; show the characteristics and action of the leg muscles.

If the specimen is a female, remove nearly all the eggs and throw them away. Insert a blowpipe in the glottis and partly inflate the lungs. Wash the specimens thoroughly to remove all traces of blood and cover them with water in a dissecting pan.

If the specimens are needed on successive days, they should be wrapped in a wet cloth immediately after the class work of each day and kept in a cold place. Use only specimens that are fresh.

1. Make an outline drawing, natural size (or twice natural size if the frogs are small), of the ventral view of the head and trunk regions of a dissected specimen, together with the base of each of the four appendages, and *draw nothing else until directed to do so.*
2. The *heart* is a cone-shaped body midway between the arms. Draw the heart to show its position, shape, and relative size.
3. On either side of the heart are the *lungs.* Stretch one of them a little by pulling on it; then letting it go.
    a. State whether or not the lungs are elastic. Are they hollow or solid? Of what advantage are these two characteristics of structure?
    b. What is the color of the lungs? State whether or not you find tiny blood vessels on the surface.

c. Draw the lungs in position to show their situation, size, and shape.
4. On the frog's right side, and behind the heart and lungs is the reddish, several-lobed *liver*. Lay the liver over to one side and find between the lobes on the underside a thin-walled, green sac, the *bile sac* (or *gall bladder*). Sketch in your drawing the liver to show it in this position together with the gall bladder.
5. On the frog's left side and under the liver in its natural position is a whitish, oblong body, which narrows at its hinder end. This body is the *stomach*. Push the handle of the dissecting needle down the gullet into the stomach.
    a. Tell what you have just done.
    b. What organ does the handle enter?
    c. Push the stomach to the frog's left and draw it in this position to show its shape and relative size.
6. Extending from the stomach is a tubular structure of considerable length, the *small intestine*. At the lower end of the small intestine the tube becomes larger and then disappears between the two thighs. This last part of the tube is called the *cloaca* or *large intestine*.
    Draw the small and large intestines.
7. Between the stomach and the first loop of the small intestine is a thin pink body, the *pancreas*, which is a very important digestive gland.
    Draw the pancreas.
8. Label heart, lungs, liver, stomach, small intestine, large intestine, bile sac, pancreas.
9. Push the small intestine to one side and find two red bodies on either side of the spinal column. These bodies are the kidneys. The kidneys remove the *nitrogenous waste* (*urea*) from the blood.
    Make a sketch of the kidneys twice the natural size.

FROGS AND THEIR RELATIVES

82. Habits of frogs. — There are many kinds of frogs, differing from one another considerably in size and color; but all frogs live in places where water is more or less abundant. Frogs are often found either on the banks of ponds and streams, or floating on the surface of the water with only the tip of the nose above water (Fig. 78). In color they usually resemble their surroundings rather closely, and so secure a certain degree of protection from fishes, snakes, birds, and

FIG. 78. — Frogs in their habitat. Four frogs are shown; in the middle of the picture a black snake is preparing to seize the frog. (Part of an exhibit at American Museum of Natural History.)

man, which are their more common enemies. When pursued, they quickly disappear beneath the water and often bury themselves in the mud at the bottom until the need of air compels them to return to the surface. Late in the autumn they burrow in the mud and remain there until the following spring. The more or less pointed snout of the frog, its slippery skin, its long, muscular legs, and its webbed feet all adapt the animal for rapid swimming through the water.

83. Food, food-getting, and digestion. — Frogs feed upon insects, fish, and other frogs, and even birds have been found in their stomachs. Insects are caught by the aid of the slimy tongue, the tip of which can be quickly thrust out of the mouth and then drawn back again with the insect adhering to it (Fig. 79). The tiny teeth that are found on the upper jaw and the two large teeth in the roof of the mouth are useful only in preventing the escape of the prey from the mouth. Hence the food is swallowed without being chewed, and after passing down the short gullet it enters the tubular stomach (Fig. 80), where it is partially digested by ferments (P. B., 53) secreted by certain cells found in the lining of this organ.

FIG. 79. — The method by which a frog secures insects.

When the food leaves the stomach, it enters the coiled small intestine where the process of digestion is continued by the bile secreted in the liver and the pancreatic juice prepared in the pancreas.[1] As the digested food slowly moves along the small intestine, it is absorbed by the capillaries (**84**) in the walls of this tube and so may be carried by the blood to the various cells of which the body is composed. Digestion not only prepares the food for absorption, but as in plants (P. B., 51) or in the fish (**98**), makes it ready to be used in the cells either for growth and repair or for the production of energy.

84. Blood and circulation. — The blood of the frog, when examined under the microscope, is seen to consist

[1] Both of these digestive fluids are carried to the intestine by ducts.

FROGS AND THEIR RELATIVES 111

of two kinds of cells (the *red* and *white corpuscles*) which are floating in a liquid known as *plasma*. The plasma consists largely of water and the digested foods that have been absorbed from the alimentary canal (83).

As in the fish, the circulatory system consists of the heart and three kinds of blood vessels; namely, arteries, veins, and

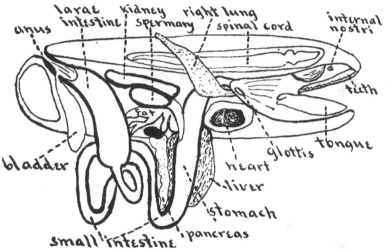

FIG. 80. — Internal organs of the frog.

capillaries. The heart is located in the body cavity just back of the head and consists of two auricles and a ventricle (Fig. 81), instead of a single auricle and ventricle as in the fish (Fig. 100). As might be expected, this makes necessary other differences in the circulatory system of the frog. In the fish we shall see that there is only one stream of blood flowing into the heart, while in the frog there are two. One stream enters the heart from the various organs of the body which the blood has supplied with food and oxygen, and from which the blood has received carbon dioxid. The second blood stream comes from the lungs where the blood has given off the carbon dioxid and received a fresh supply of oxygen. The right auricle receives the blood brought from the body in three large veins, while two small veins carry into the left auricle the blood from the lungs.

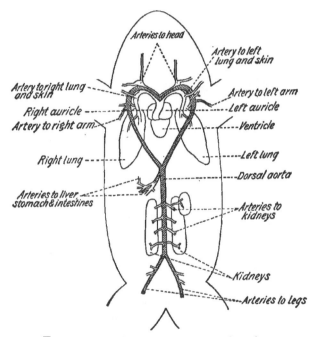

FIG. 81.—Arteries in the circulation of the frog.

The blood from both auricles now flows into the single ventricle, which then contracts and pumps the blood into a large artery. Certain branches carry the blood having the larger amounts of oxygen (*i.e.* the blood from the lungs) to the head, trunk, legs, and other organs of the body, while other branches carry the blood just received from the body, with its larger amount of carbon dioxid, to the lungs and the skin.

In the capillaries which connect the arteries and veins in every part of the body (Fig. 82) all the changes in the composition of the blood take place, since their thin walls permit the food materials and oxygen to enter the cells and the wastes from the cells to enter the blood.[1] The capillaries in the lungs likewise permit the interchange of oxygen and carbon dioxid.

FIG. 82. — Network of capillaries connecting an artery and a vein.

FIG. 83. — Capillaries in web of frog's foot.

85. Respiration and the liberation of energy. — The walls of the frog's lungs contain a network of capillaries, and in these thin-walled tubes the red corpuscles absorb the oxygen that is forced into these sacs by the upward movement of the floor of the mouth. As we have seen, the blood with a fresh supply of oxygen flows from capillaries of the lungs into

[1] A tadpole's tail is excellent for demonstrating the blood current. Wrap a tadpole in wet cloth or cotton and support it so that the tail can be placed between two glass slides on the stage of the microscope. The space between the two slides should be kept filled with water. The movement of the corpuscles through the margin of the tail should be examined with the low power of the microscope (Fig. 83).

veins and so finally into the left auricle and thence into the ventricle. Here it tends to become somewhat mixed with the blood from the right auricle which has just returned from the body. However, the structure of the heart and the arteries is such that the blood that has come from the lungs with a larger supply of oxygen is sent out by arteries to all parts of the body.

In the capillaries the oxygen is absorbed by the cells. Oxidation of the food and protoplasm takes place and energy is thereby released, which enables the frog to carry on locomotion, secure its food, and perform all its destined tasks. The carbon dioxid and other wastes produced by oxidation pass through the capillary walls into the blood and, as we have seen, are carried back to the heart and then to the lungs, where carbon dioxid is excreted. Other wastes are excreted by the kidneys.

The skin of the frog is likewise permeated by a network of capillaries so that it acts as do the gills of fishes in absorbing oxygen from the water and in giving off carbon dioxid. While the frog is buried in the mud during the winter it breathes entirely through the skin. So much does the frog depend on the skin as a breathing organ that even in summer, if the skin becomes dry so that air cannot be absorbed, the frog dies.

86. Reproduction and life history. — In the animals studied thus far we have found special organs devoted to the process of reproduction, namely, ovaries for egg production in the female and in the male spermaries that form the sperm-cells. Before the egg-cells can develop into embryos each must be fertilized by a sperm-cell. All the facts we have just stated apply equally well to the frog.

Frogs' eggs are deposited in springtime in masses that

FROGS AND THEIR RELATIVES

A, eggs before they are laid

B, eggs after they are laid

C, egg containing young tadpole

D, young tadpoles attaching themselves to a plant

E, young tadpole with external gills

F, young tadpole with internal gills

G, young tadpole with hind legs

H, tadpole with webbed feet

I, tadpole with legs and arms

J, young frog

Fig. 84. — Life history of frog.

A, one-celled stage

B, two-celled stage

C, four-celled stage

D, eight-celled stage

E, sixteen-celled stage

F, thirty-two celled stage

G, sixty-four celled stage

H, many-celled stage

Fig. 85. — Cell division in a frog's egg.

float on the surface of the water [1] (Fig. 84, B). Each fertilized egg is a small sphere, black on its upper surface and white beneath, and inclosed in a gelatinous covering. The warmth of the sun causes the one-celled egg to divide vertically in half to form the two-celled stage (Fig. 85, B) and the process of division continues until the egg consists of many cells (Fig. 85, H). Food for the development of the embryo is stored in the egg.

The many cells of the egg gradually become different in character and so form the various organs of the embryo (Fig. 86). Soon after hatching, the young of the frog, known as *tadpoles*, secure their food by sucking in tiny water plants found on the surface of plants and stones (Fig. 84, D). Tadpoles resemble fishes in having gills for breathing, a heart with two chambers instead of three, and a tail for locomotion. At first the gills are on the outside of the body (Fig. 84, E), but later four pairs of internal gills are formed, and the external gills are absorbed. The animal increases in size, the hind legs appear, and the arms are formed beneath the skin. Meanwhile the lungs are being developed, the heart becomes three chambered, the legs grow larger, arms appear, and finally the gills and the tail are completely absorbed. The tadpole now leaves the water, since it is an air-breathing animal. This succession of changes after hatching from the egg is known as a *metamorphosis*.

FIG. 86. — Embryo of frog.

87. Relatives of the frog. — Much like the frog in structure and life history is the common garden toad. Toads, however, in their

---

[1] If possible eggs in different stages of segmentation should be secured, preserved in 5 per cent formalin, and used for demonstration.

adult stage cover themselves more or less with dirt in the daytime, and come out at night to feed upon insects, which constitute their sole food. Instead of having a smooth, slimy skin, as does the frog, a toad's skin (Fig. 87) is dry and covered with elevations commonly known as "warts." These elevations contain cells which secrete an irritating substance that protects the toad from animals

Fig. 87.—The toad. Note its resemblance to its surroundings, whereby it is likely to be protected.

that would prey upon it. There is no foundation, however, for the popular notion that the warts of human beings are ever caused either by toads or frogs.

In springtime toads seek the water in which to breed. The eggs, covered with a gelatinous substance are laid in long strings instead of in masses, as was the case with frogs. The development and life history of the toad is much the same as in the case of the frog. As soon as metamorphosis is complete, the little toads leave the water and often are found considerable distances away from water.

Less like the frog, at least in its adult stage, are the salamanders and newts (Fig. 88). These are found in damp places or in

Fig. 88. — The newt.

water and are often called "lizards," by those who do not know that a lizard has scales, claws on its feet, and breathes throughout its life by means of lungs. Some of the relatives of the frog, even after they have developed lungs, retain gills throughout their life (Fig. 89).

Fig. 89. — A mature amphibian (Necturus) with external gills.

Because of the ability of the animals described in this chapter to live both on the land and in the water, they are called the *amphibia*, from Greek *amphi* = both + *bios* = life.

88. Economic importance of the amphibia. — None of the amphibia, so far as is known, are harmful to man. On the contrary, all of them are more or less useful because of the insects that they devour. This is especially true of the garden toad. It has been estimated by one author that a toad in a garden is worth nearly twenty dollars a year on account of the cutworms and other injurious insects that it destroys. "In France the gardeners even buy toads to aid them in keeping obnoxious insects under control."—HEGNER's "College Zoölogy."

Frogs, in addition to their value as insect destroyers, are also of some value to man as food. It is said that in the United States about $50,000 is obtained annually by the frog hunters for their catch, and frog farms are now profitably maintained in several states. Frogs are also used as fish bait.

CHAPTER IV

FISHES

89. What is a fish?—"A fish is a backboned animal which lives in the water and cannot ever live very long anywhere else. Its ancestors have always dwelt in water, and likely its descendants will forever follow their example. So, as the

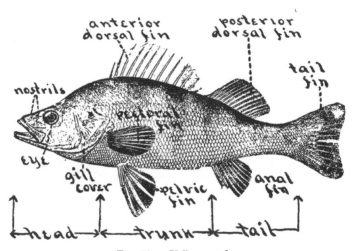

Fig. 90.—Yellow perch.

water is a region very different from the fields or the woods, a fish in form and structure must be quite unlike all the beasts and birds that walk or creep or fly above ground, breathing air, and being fitted to live in it. There are a great many kinds of animals called fishes, but in this all of them agree: all have some sort of a backbone, all of them breathe their life long by means of gills, and none have fingers or toes with which to creep about on land." [1]

90. The regions and appendages of a yellow perch. — Study Figure 90 and notice that the body of the yellow perch is divided into three regions; namely, *head, trunk,* and *tail.* Unlike the body of many animals, no neck is present, and the head, therefore, is joined directly to the trunk. The line of union of head and trunk is the posterior [2] margin of movable flaps, called the *gill covers,* on the sides of the head. Just behind or posterior to the gill cover on each side of the trunk of the fish is a paddle-like organ called the *pectoral fin.* On the ventral surface, below the pectoral fins, is a second pair which are known as the *pelvic fins.* The pectoral and pelvic fins are together known as the *paired fins* of the fish. Besides these this animal has several *unpaired fins,* which we shall now locate. On the dorsal surface notice two *dorsal fins,* one behind the other, which project upward. Below the posterior dorsal fin, on the ventral surface, is another single fin called the *anal fin.* The tail region is considered to begin just in front of the anal fin, since in the fish the body cavity that contains the important organs of digestion, circulation, and reproduction ends at this point (Fig. 98). The anal fin, therefore, and also most of the posterior dorsal fin, are attached to the tail region. At the posterior end of this third region is the broad forked *tail fin.*

91. Regions and appendages of a goldfish. — Laboratory study.

[1] Jordan's "Guide to the Study of Fishes," Vol. I, p. 3.
[2] The meaning of each of these terms is explained in **6**.

Materials: A living goldfish in a battery jar for each two students. Goldfish may be kept indefinitely in a glass jar with green water plants; the latter supply the fish with food and oxygen. Perch, and if possible the heads of large fishes like the cod, should be obtained, preserved in formalin (5 per cent), and then thoroughly washed in running water for twenty-four hours before they are used; material treated in this way loses its fishy smell, and may be kept in the formalin solution year after year. A fish skeleton is also needed for demonstration. The Jung charts of the external and internal structure of the perch are useful.

Observe a living goldfish and compare it with Figure 90.
1. Name the regions of its body and state, with reference to gill cover and fins, where each region begins and ends.
2. Name and locate all the organs you find on the head.
3. What paired and what unpaired fins are found on the trunk? Using the terms anterior, posterior, dorsal, ventral, median, and lateral, locate each of these fins.
4. Name and locate the fins attached to the tail region of the body.
5. Make an outline sketch about five inches long of the side view of a living goldfish to show the shape and relative size of the three regions, the position and shape of the organs of the head and of the various kinds of fins. Label the regions and the organs that you have drawn, in a manner similar to Figure 90.

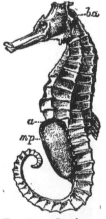

FIG. 91.—Sea horse.

92. Some differences in the form of fishes. — One can usually tell whether or not an animal is a fish; but in some cases this is

FISHES 123

extremely difficult. Who would think, for instance, that such animals as the sea horse (Fig. 91) and the pipefish (Fig. 92) would be

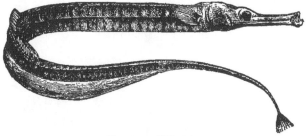

Fig. 92. — Pipefish.

classed with the perch and goldfish? Yet such is the case, since careful study has shown that these forms have all the characteristics mentioned in 89.

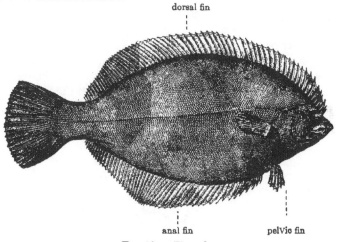

Fig. 93. — Flounder.

It is evident that the goldfish and perch have bodies that are considerably longer than they are wide or deep, and this is true of most of the common fishes. In the group of fishes known as the eels, this elongation is so marked that they look more like snakes than they do like fishes. But the eels are not the only fishes that show a striking development in one dimension. The flounders, for example (Fig. 93), exhibit a notable growth in a dorso-ventral direc-

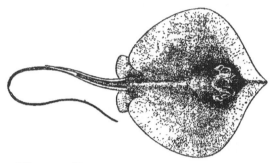

Fig. 94. Sting ray. (Jordan and Evermann. Courtesy of Doubleday, Page & Co.)

tion. So far has this been carried that the fish is unable to retain a vertical position, and consequently lies on one of its sides. The eyes, which, in very young flounders, are situated like those of the goldfish, on either side of the head, by a twisting of the bones of the

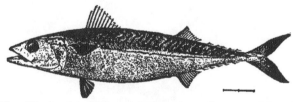

Fig. 95. — Mackerel. (Jordan and Evermann. Courtesy of Doubleday, Page & Co.)

skull, both come to lie on the same side of the head. Otherwise, as may be seen, one of the eyes would rest on the sand or mud, when the animal is on the sea bottom. Fishes like the skates and sting rays (Fig. 94) have also a much flattened body, but these animals have

FISHES 125

attained this condition by growth from side to side, instead of dorso-ventrally.

93. Some differences in the fins of fishes. — We have seen that the goldfish has one dorsal fin, the perch two, and that both fishes have a single anal fin. A glance at Figure 108 will show that the cod fish has three dorsal fins and two anal fins. Dorsal and anal fins vary not only in number, but in extent. In some fishes they are very short, as in the mackerel (Fig. 95), while in the flounder (Fig. 93) these fins extend nearly the whole length of the dorsal and ventral surfaces.

Most common fishes possess both pectoral and pelvic fins, but in the eels (Fig. 96) the pelvic fins are entirely wanting and the pec-

FIG. 96. — Eel. (Jordan and Evermann. Courtesy of Doubleday, Page & Co.)

toral fins are very small. The paired fins vary in position as well. In the perch, for example (Fig. 90) the pelvic fins are immediately below the pectorals, while in the cod (Fig. 108) they are anterior to the pectoral fins, and in the salmon (Fig. 107) they are even farther back on the body than in the goldfish.

94. Adaptations for swimming. — Laboratory study.

1. Carefully watch for a time a goldfish when it is swimming around in a large battery jar or aquarium.
    a. Which of the three regions of the body is principally used in pushing the animal forward?
    b. Describe the movements of this body region.
2. Which of the paired fins are used in swimming? De-

scribe their movements. State whether or not you see the fish swim backward.
3. If the goldfish strikes backward with the fins against the water, would the fish tend to move forward or backward?
4. Since the goldfish moves the fins both backward and forward in the water, in which direction must it strike the harder and more swiftly if it wishes to swim forward?
5. (Optional.) Suppose the fish strikes backward harder with the fins on the right side than it does with those on the left side, how would the direction of its motion be affected?
6. (Optional demonstration.) Place the largest goldfish you can get, in a sink or other large receptacle full of water. Get the fish to swim continuously and rapidly, but not so rapidly that the pupils are unable to see the paired fins.
   a. What have you seen that leads you to think that the goldfish does not use the paired fins in rapid swimming?
   b. What parts does the animal use to drive itself rapidly through the water?
7. Why are the broad, flat surfaces of the fins of advantage to the fish in swimming?
8. Study the anterior, dorsal fin of a perch or other fish. Notice that it is composed of stiff *fin rays* and of thin *connecting membrane*. Alternately spread out and close the fin, and bend each of the materials of which it is composed. Now describe the structure of this fin.
9. Examine carefully each of the fins of a goldfish.
   a. State whether or not each consists of fin rays and connecting material.
   b. What disadvantage to a goldfish in swimming would result from the absence of the rays in a fin?
   c. State the relative difference in the size of a fin when it is spread open and when it is closed.
   d. What would be the disadvantage if the open fin had no connecting membrane?

95. The locomotion of fishes. — Many fishes, like the goldfish and perch, are able to maintain a given position in the water while at rest. This is made possible by means of an internal organ known as the *swim bladder* (Fig. 98). The swim bladder may be compressed, permitting the fish to sink, or it may be expanded, causing the animal to rise. Since, therefore, the fish is poised in a liquid medium, it is only necessary to overcome the resistance of the water about it in order to move in any given direction. This resistance is more easily overcome, first, because the head is somewhat pointed like the prow of a boat, secondly, because the overlapping *scales* point backward, and third, because the whole body is covered with a slimy *mucus*.

One who is at all familiar with a canoe knows that it is impossible to propel it by the use of a slender rod. One must have a paddle with the lower end broad and flat so that sufficient force may be exerted against the water to propel the canoe. Now, in swimming, the fins of a fish act more or less like paddles. Their broad, flat surfaces press against such an amount of water that the fish is enabled to exert enough force to push its body in any desired direction.

If one watches a goldfish swimming slowly about in an aquarium, one would think that the paired fins, especially the pectoral fins, were the important swimming organs. But careful experiments have shown that this is not the case. When the goldfish has occasion to move more rapidly, the paired fins are not used at all, but are pressed close to the sides, the body being driven through the water by the movement of the tail and tail fin. The paired fins, together with the dorsal and anal fins, seem to be used principally in steering the fish. The energy necessary for swimming is developed in the powerful muscles of the tail and trunk.

96. Adaptations for food getting. — Laboratory study.
1. Open the jaws of a fresh or of a preserved fish. (Fish of large size, *e.g.* cod, should be used if possible, the jaws being held wide open with pieces of wood.) Look for teeth on the jaws and the roof of the mouth.
    a. State the location of the teeth and give some idea of their number.
    b. Rub the fingers gently back and forth over the teeth. Do they point backward or forward? How do you know?
    Describe any other characteristics of the teeth.
    c. Of what use would the teeth be in catching other fish for food?
    d. Why would the shape of the teeth make them of no use in grinding food?
2. Drop some fish food into a jar containing living goldfish.[1] Describe all the movements that you see the fish make while feeding.

97. Food and food getting among fishes. — Unlike plants, fishes cannot make their food from materials found in the water, air, and soil, but must secure it ready-made from plants or other animals. The goldfish, for example, depends largely on vegetable food, while the cod[2] and the perch for the most part feed upon other animals smaller than themselves. Since these fishes must catch and hold their prey, their jaws are provided with many sharp teeth that point backward, and so prevent the escape of any active

---

[1] If none of the fish eat readily, this experiment should be deferred.
[2] "The cod is omnivorous, and feeds upon various kinds of animals, including crustaceans, molluscs, and small fishes, and even browses upon Irish moss and other aquatic vegetation. All sorts of things have been found in cods' stomachs, such as scissors, oil cans, finger rings, rocks, potato parings, corn cobs, rubber dolls, pieces of clothing, the heel of a boot, as well as other new and rare specimens of mollusks and crustacea." — JORDAN and EVERMANN, "American Food and Game Fishes."

animal which may have been caught. The cod, as you may have seen, has teeth in the roof of the mouth and in the throat in addition to those found on the jaws, thus making more secure its hold upon the unfortunate denizen of the deep that it has seized.

Certain fishes depend on minute forms of plants and animals, and therefore some means is needed by which the water taken in with the food may be gotten rid of while at the same time the food is retained. Hence, fishes are provided with a straining apparatus which permits the water to escape when the mouth is closed, and retains within the mouth the minute forms of life that it has secured. Of this adaptation for food getting, we shall learn more in our study of the gills.

Most of the fishes that prey on other animals secure their victims by dint of their speed; but one form of fish, called

FIG. 97. — Deep sea angler.

the "deep sea angler" (Fig. 97), has upon the dorsal part of the head a bulbous projection, the tip end of which is luminous. This bright light attracts other fishes, and when they approach near enough, the "angler" makes a quick dash, closes its big jaws upon the too curious individual, and so

130   ANIMAL BIOLOGY

secures food. But whatever a fish feeds upon, and however it secures its food, it is evident that plants and other animals must furnish the food substance required to make living matter, and so provide for growth and repair of the cells, and also furnish the fuel needed to develop the energy necessary for the various activities of the fish.

98. Digestion and digestive organs. — We have seen in plants (**P. B.**, **63, 70, 74**) that digestion may take place in any living cell where food is stored or manufactured. Hence plants have no special part devoted to digestion. In fishes however, it is quite different, since a portion of the body, known as a digestive system, is devoted to preparing the food for absorption and use. This digestive system consists of a food tube known as the *alimentary canal* and certain masses of cells known as *digestive glands*.

When the fish swallows food, this passes from the mouth cavity into a short tube, called the *gullet*, and thence into a

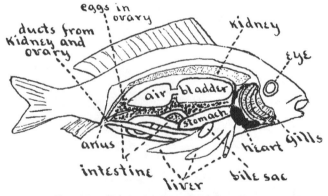

FIG. 98. — Internal organs of a fish. (Carp.)

comparatively long wide *stomach* (Fig. 98), which in the carp extends half the length of the body cavity. From the stomach extends the *small intestine*, which turns upon itself several times, thus forming a coil, the posterior end of which finally opens to the exterior just in front of the anal fin.

In the inner lining of the stomach and intestine are special cells which make up digestive glands. These have the power to manufacture digestive ferments (**P. B.**, **53**), which are forced out into the alimentary canal when food is present. As in plants, these ferments dissolve the foods and make them ready for use in the body. In addition to the digestive glands in the lining of the alimentary canal there are glands outside the digestive tube. One of these is the *liver* (Fig. 98), which secretes bile. This is carried to the intestines by a tube called the bile duct. In the liver is a sac (*bile sac*) (Fig. 98) which holds any excess of bile. When the food has been digested it is absorbed by thin-walled blood vessels found in the lining of the alimentary canal, and so passes into the blood to be distributed around the body.

99. Blood and circulation. — Instead of ducts and sieve tubes (**P. B.**, Figs. 14, 15, 16) as in the seed plants we studied, the fish has *blood vessels* to distribute digested foods to various parts of the body. In addition to these the fish possesses a *heart* (Fig. 99), which aids in pumping or forcing the blood through blood vessels, thus keeping it in constant motion. The blood vessels are of three kinds; namely, *arteries, capillaries, and veins*. The arteries have muscular and elastic walls which contract and so aid the heart in forcing the blood along its course. The arteries always carry the blood away from the heart, and they subdivide into smaller and smaller tubes. At the ends of the smallest arteries are tiny, short, thin-walled blood vessels, known as capillaries. Capillaries permit the digested food to osmose through their walls into the adjacent cells, and, in turn, absorb waste matters from the cells.

132 ANIMAL BIOLOGY

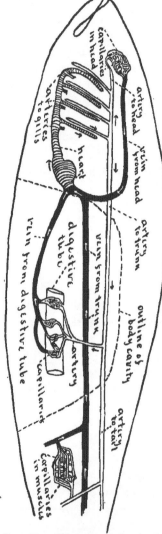

FIG. 99.—Diagram of the circulation of a fish.

The blood passes from the capillaries into the veins,[1] which are thinner-walled than the arteries. These veins carry the blood back to the heart. The heart (Fig. 100) consists of two principal parts; a thin-walled *auricle* which receives blood from the veins, and a thick-walled, muscular portion called the *ventricle*, which forces the blood out into the arteries.

100. Adaptations for breathing. — Laboratory study.

1. Raise the gill covers of a preserved fish and find the gills. Carefully separate the gills with the forceps. How many gills are present on each side?
2. The openings between the gills are called *gill clefts*. Gently push a thin strip of wood or the forceps through one of the gill clefts as far as you can.
    a. In what cavity do the forceps or strip of wood appear?

[1] This is true of all organs of the fish, excepting the gills. See 101.

b. Describe the situation of the gills with reference to the mouth cavity.
3. Hold the preserved fish with its mouth upward and carefully pour water into the mouth opening. Where does the water come out?
4. Place a gill that has been removed from a fish (a salmon if possible) in a watch glass and cover it with water. Find the following parts : (1)a soft part made up of slender divisions called *gill filaments*. (2) a curved part to which the gill filaments are attached, known as the *gill arch*, and (3) projections on the side opposite the gill filaments which are known as *gill teeth* or *rakers*.
   a. Is the gill arch relatively hard or soft compared to the filaments?
   b. Are the gill teeth or rakers relatively hard or soft?
   c. Look again at the perch and state whether the rakers are found on the side of the arch nearest to the mouth cavity or on the opposite side.
   What is the use of the gill teeth when the fish takes in a mouthful of water containing food, and does not wish to swallow the water?
   d. Make a sketch (about four inches long) of a gill to show the shape of the whole and the structure of a small portion. Label gill arch, gill filaments, gill rakers.
5. The gill filaments contain thin-walled blood vessels (capillaries) which are separated from the water by a thin membrane. The heart forces the blood into certain arteries that carry it to the capillaries in the gills and thence blood passes back to the body through another set of arteries. (Fig. 100.) Bearing in mind that breathing is essentially the same in animals as in plants (**P. B.**, 82), —
   a. What gas will the blood bring from the body to be given off in the gills in the process of breathing?
   b. What gas is taken up by the blood in the gills to be carried around through the body?
   c. How are the gill filaments (as stated above) fitted by structure to permit this interchange of gases?

d. How are the delicate gill filaments protected from injury?
6. If the same water remained on the gills for some time, what changes in the relative amount of oxygen and carbon dioxid in the water would occur? Why, then, is it necessary that a current of water should pass over the gills?

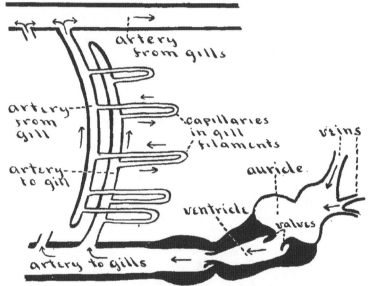

FIG. 100. — Diagram of the circulation of blood in the gill of a fish.

7. Describe the movements of the jaws and gill covers of a living goldfish when it is breathing. (If the fish has been in a jar of water without green plants for some time, these movements will be more pronounced.)
8. Watch the fish as it opens its mouth.
   a. Is the size of the mouth cavity now greater or less than it was when closed?
   b. Why does the water now enter the mouth? (The inward movement of the water may be demonstrated

FISHES

more easily if some powdered carmine is stirred into the water.)
  c What will the incoming current of water bring to the gill filaments?
9. Watch the fish as it closes its mouth.
  a. Is the size of the mouth cavity now greater or less than it was before?
  b. Why do the gill covers now open?
  c. What will the current of water carry away from the gill filaments?

101. Respiration and the production of energy. — We have just seen that when the goldfish takes in a mouthful of water and then closes its mouth, the water is forced over the gills, thus bringing oxygen to the filaments. The capillaries in the filaments absorb the oxygen, and the blood then passes on into other arteries which carry it all over the body of the fish. In the capillaries at the ends of the smallest arteries the oxygen passes into the cells as does the food. Now what becomes of the oxygen?

As in plants (**P. B.**, 80), the oxygen unites with elements in the foods and in the protoplasm of the cells and produces oxidation and liberation of energy, which gives the fish the power to contract its muscles and so to push against the water with its tail and tail fin, thus propelling the animal in any direction, or to open its jaws and shut them on another fish, thus securing food. In fact, all the work that the fish performs is made possible through the burning of its foods or protoplasm by the oxygen.

Since the proteins, fats, carbohydrates, and protoplasm all contain carbon, when these are oxidized, carbon dioxid (CO_2) is formed as one of the waste substances. All the waste substances pass out of the cells, through the walls of the capillaries, into the blood, which passes on into the veins and back to the heart. The heart contracts and drives the blood loaded with carbon dioxid out into the arteries, which carry it to the capillaries of the gills (Fig. 100). Here the waste matters pass out into the water, which is then forced out by the closing of the mouth past the gill covers.

102. Adaptations for sensation. (Optional.)
1. Study the eye of a goldfish.
    a. Describe its position, shape, and size relative to that of the head.
    b. Notice that the eye consists of a black center (the *pupil*) through which light enters the eye, and a colored *iris*. Add these features to the drawing of the goldfish (**91,** 5), and label each.
2. The nostrils lie in front of the eyes, and as they are small, a preserved fish head may help in locating them. (In the perch there are two on each side.)
    a. Show in your drawing the position, shape, and size of the nostril of one side and label.
    b. Gently probe the nostril of a preserved fish with a stiff bristle.
        (1) Do the nostrils open into the mouth or not?
        (2) Could the nostrils be used in breathing? Give reason for your answer.
        (3) Bearing in mind the common uses of nostrils of higher animals, state which of these is the probable function of the nostrils of a fish.

103. Senses of fishes. — Fishes are said to possess keen sight. The eyes, however, except in rare cases, are only fitted for seeing while in the water. These organs have no eyelids, so the fish always seems to be wide awake. The sense of smell is located in the nostrils, and since these do not open into the mouth cavity, this is the only function of the nostrils. The taste sense is said to be located in the outer skin. The fish has no external ears; it has, however, internal ears, but these are supposed to serve as balancing organs, rather than as organs of hearing. Fishes from which these internal ears have been removed are unable to maintain their equilibrium.

Some fishes have special organs that serve as tactile organs such as are found on the under side of the head of a cod (Fig. 108) and also on the head of bullheads (Fig. 101). Along each side of the body and tail of fishes is a series of little openings or pores which form what is known as the *lateral line* (Fig. 108). These organs are supposed to be principally organs of touch.

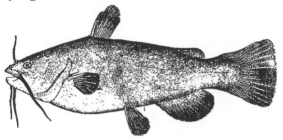

FIG. 101. — Bullhead. (Goode.)

104. Reproduction and life history. — The flowers of seed plants are devoted to the production of seeds which, in turn, produce new plants of the same kind (**P. B., 83**). Likewise in fishes there are special organs the sole function of which is the production of new individuals. The organs of fishes which may be said to correspond in function to the stamens and pistils of flowers are the *ovaries* (Fig. 98) and *spermaries*. In the ovaries are produced many *egg-cells*, and the mass of eggs in the ovary of a fish is often called the *roe*. In order that an egg may develop it must first be *fertilized* by a *sperm-cell* from the spermary of a male fish. This process usually occurs in the water after the ripe eggs and sperm-cells have been extruded from the ovaries and spermaries of the parent fishes.

You will recall the fact that the pollen tube containing a sperm-nucleus makes its way into an ovule and that the

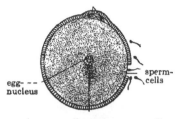
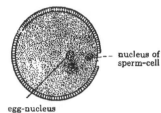

A, sperm-cell entering an egg-cell

B, sperm-nucleus approaching the egg-nucleus

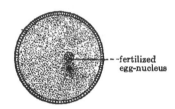

C, sperm-nucleus and egg-nucleus uniting

D, fertilized egg-nucleus

FIG. 102. Fertilization of an egg.

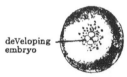

A, four-celled stage of embryo

B, many-celled stage of embryo

C, embryo more fully developed

D, young fish with yolk still attached

FIG. 103. — Development of a fish egg.

sperm-nucleus is forced into the ovule and unites with the egg-nucleus; this is the process known as *fertilization* (**P. B.**, 91). In the case of fishes the sperm-cells swim to the eggs, and then force their way into the egg (Fig. 102, *A*).

Fig. 104. — Nest of stickleback. Above, male entering nest with eggs; below, male depositing sperm-cells.

The nucleus of the sperm- and egg-cells then unite just as in plants (Fig. 102, *B, C, D*). The egg nucleus thus fertilized first divides, and then the cell body, and thus are formed two cells. Each of these cells in turn divides, and so four cells are produced (Fig. 103, *A, B*). The process of division continues until a many-celled organism is developed.

As the cells increase in number, they become different in character and form the various organs of the body. When the little fish first hatches, and begins to swim about, it often has attached to it some of the food substance (yolk) stored in the egg (Fig. 103, *D*). After this is used up, the young fish must secure its own food.

Most fishes do not take any care of their eggs or young, and in some cases the parents die soon after the eggs are laid and fertilized. In the case of the stickleback, however, the male fish makes a nest (Fig. 104) in which the females deposit their eggs. The male then extrudes sperm over the eggs. The male stays about the nest and guards the eggs and also the young sticklebacks when they hatch out.

FIG. 105. — Artificial fertilization of eggs. (Coleman.)

105. Artificial propagation of fishes. — Since, as we have said, most kinds of fishes give no attention to eggs or young, enormous numbers of both eggs and young are eaten by other fishes; hence, only a small proportion come to maturity. For example, while a codfish lays 8,000,000 eggs, only about two of these eggs on the average come to maturity. Hence, in order to increase to any considerable extent the number of fishes, the eggs are artificially hatched. That is, the fish

FISHES 141

are caught when the eggs are ripe and the eggs are gently squeezed from the ovaries into the water (Fig. 105). Then some of the sperm-cells are similarly squeezed from the male fish and mixed with the eggs. This provides for fertilizing most of the eggs, which would probably not occur in nature. Special apparatus is devised for keeping the eggs supplied with fresh water until they hatch (Fig. 106). When the

FIG. 106. — Interior of fish hatchery.

young are old enough they are fed for a time, then the *young fry*, are set free in the waters where more fish are desired. Millions of young fish are every year distributed by the government all over the United States to be placed in ponds, rivers, and lakes where the supply is deficient, or in the ocean along the shore.

106. Economic importance of fishes. — From very ancient times fishes have formed a considerable part of the food of peoples that lived near bodies of water. The importance to man of fishes as a source of food can scarcely be overestimated. Unlike domestic animals, the fishes grow to maturity without any care on the part of man. The fisherman has only to provide the means to gather his harvest, while the herdsman must care for his flocks and herds the year round. Thus we see why fish are cheaper than other forms of flesh food.

While fish are most important to man as food, they have other uses. Thus, for instance, the menhaden are caught scarcely at all for food, but for the large quantities of oil extracted from them. The remainder of their bodies is used as fertilizer. It is estimated that about 3,000,000 gallons of oil and 1,000,000 tons of scrap, with a total value of $2,500,000 is obtained annually by American fishermen from this kind of fish. The oil extracted from the livers of cod forms a valuable food preparation for invalids, since it is said to be more easily absorbed and oxidized than any other known fat.

The great importance of fishes, however, is due to the fact that they furnish a cheap and wholesome food. Nearly all the parts of a fish are thus used. Not only is the flesh eaten, but also the eggs (roe). The swim bladders, too, of many fishes are made into isinglass which yields the highest grade of gelatin.[1] Fish are eaten not only in a fresh condition, but are also prepared in various ways. Among these methods of preservation are drying, smoking, pickling, and canning. Two of the more important fisheries are those of the salmon and the cod.

107. The salmon. — The salmon (Fig. 107) is doubtless the most important food fish of the world, and the Pacific salmon completely outclasses all other forms. The Atlantic

---

[1] See article on isinglass in Cyclopedia.

salmon was once very abundant, but is gradually diminishing in numbers for reasons that will be mentioned later (**110**). "The salmon were made for the millions. The Siwash Indian eats them fresh in summer, dries them, or later on freezes them, for himself and his dogs in winter. The epicure pays for having the fresh fish shipped in ice to his table, wherever that table may happen to be. In mid-ocean, the great American canned salmon is often the best and only fish afloat. In the jungles of the Far East, in the frontier

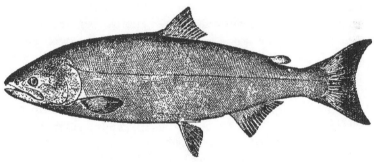

FIG. 107. — The Salmon. (Jordan and Evermann. Courtesy of Doubleday, Page & Co.)

bazaar of the enterprising Chinese trader, it 'bobs up serenely' to greet and cheer the lonesome white man who is far from home and meat markets. Even in the wilds of Borneo its name is known and respected; and he who goes beyond the last empty salmon tin, truly goes beyond the pale of civilization. The diffusion of knowledge among men is not much greater than the diffusion of canned salmon; and the farther Americans travel from home, the more they rejoice that it follows the flag.

"The common salmon of Europe, and also of Labrador and New England, was accounted a wonderful fish both for sport and for the table, until the discovery of the salmon millions of the Pacific Coast cheapened the name. To hold their place in the hearts of sportsmen, game fishes must not inhabit streams so thickly that they are crowded for room, and can be caught with pitchforks. Yet this once was true of the salmon in several streams of the Pacific Coast. The bears of Alaska grow big and fat on the salmon which they catch with the hooks that Nature gave them."[1]

The Pacific salmon are caught in the rivers that empty into the Pacific Ocean, such as the Columbia, Sacramento, and Yukon. The salmon reach their maturity in the ocean. When, however, the spawning time approaches, the salmon make their way in great numbers to the mouths of rivers like the Columbia and proceed up these streams, leaping seemingly impassable waterfalls in order to reach the headwaters. Here the sand is scooped out by the male, and the female salmon deposits her eggs and the male the fertilizing sperm-cells. The fertilized eggs are then covered with sand. The parent fish soon die; none ever reach the ocean again. After the eggs hatch, the young slowly float down the stream to the ocean to repeat the life of their parents.

It is when the fish are proceeding up the rivers that they are caught. Sometimes they are so abundant that the river seems to be choked with them. Salmon are shipped fresh in ice. Enormous quantities are also canned and smoked. The estimated value of the annual catch of Pacific salmon varies from $10,000,000 to $15,000,000.

108. The codfish. — Next in importance to the salmon, at least in the United States, is the cod (Fig 108), for which the fishermen receive about $3,000,000. Other countries engaged in the cod fisheries are Newfoundland, Canada, Nor-

[1] From Hornaday's "American Natural History."

way, Sweden, Great Britain, and France. The catch of cod for the world is estimated to be $20,000,000 annually. Codfish are found in the northern part of both the Atlantic and Pacific Oceans, but the Alaskan cod is not considered to be as fine a food fish as the Atlantic species.

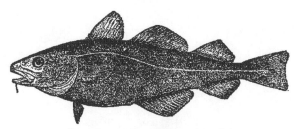

FIG. 108. — The codfish. (Goode.)

The cod is a deep water fish and is usually caught in from thirty to seventy fathoms (a fathom being six feet). Cod are caught off the coast of Newfoundland, and during the winter as far south as the Middle States. " From the earliest settlement of America the cod has been the most valuable of our Atlantic Coast fishes. Indeed the codfish of the Banks of Newfoundland was one of the principal inducements which led England to establish colonies in America, and in the records of early voyages are many references to the abundance of codfish along our shores. So important was the cod in the early history of this country that it was placed upon the colonial seal of Massachusetts, and it was also placed upon a Nova Scotian bank note, with the legend ' Success to the Fisheries.' "[1]

The average weight of large cod is said to be from twenty

[1] From Jordan and Evermann's "American Food and Game Fishes." Every student of this fish work should read Kipling's "Captains Courageous" for description of the cod fisheries on the Grand Banks.

to thirty-five pounds, depending on the locality. The average weight of small cod is twelve pounds. Jordan and Evermann state that cod weighing 75 pounds are not common,

Fig. 109. — The shad. (Goode.)

but that one was caught off the New England coast that weighed 211½ pounds.

Codfish are marketed fresh, pickled, salted, and dried. Oil and isinglass are also obtained from the cod.

Fig. 110. — The herring. (Jordan and Evermann. Courtesy of Doubleday, Page & Co.)

109. Library study of other fishes. — (Optional.)
Consult Jordan and Evermann's "American Food and Game Fishes," Hornaday's "American Natural History," and special articles in Cyclopedias or other reference books, on one or more of the following fishes: mackerel (Fig. 95), sardine, shad (Fig. 109), herring (Fig. 110), white fish, smelt, bluefish, halibut, menhaden. Write in your notebook an account of the fishes selected for study, using the following topics as a guide: —

FISHES

1. General appearance (size, color, general form).
2. Geographical distribution (that is, in what waters the fishes are found).
3. Food and feeding habits.
4. Method of capturing the fish.
5. Amount caught annually and its value in money.
6. Breeding habits and other general facts of interest.

If possible illustrate your composition with any drawings or pictures of the fish you are studying.

110. Visit to a fish market. — (Optional.)

In your notebook prepare an account of your visit to some fish market, using the following topics as a guide.

1. Location of the fish market and the name of its owner.
2. Make a list of the various kinds of fish offered for sale.
3. State the kind of fish that sells at the lowest price per pound at this time of year.
4. State the kind of fish that is most expensive per pound at this season.
5. Name the kind of fish now sold in the greatest quantity.

111. Conservation of food fishes. — A story of reckless waste similar to that recorded in regard to the destruction of our forests may be duplicated here concerning the way men have exploited our abundant natural source of food, the fishes. The Atlantic salmon, which was once "the salmon," is now of comparatively little commercial importance. "Salmon were marvelously abundant in colonial days. It is stated that the epicurean apprentices of Connecticut would eat salmon no oftener than twice a week. . . . There can be no doubt that one hundred years ago salmon fishing was an important food resource in southern New England. . . . But at the beginning of this century salmon began rapidly to diminish. Mitchill stated in 1814 that in former days the supply to the New York market usually came from the Connecticut, but of late years from the Kennebec, covered with ice. Rev. David Dudley Field, writing in 1819, states that salmon had scarcely been seen in the Connecticut for fifteen or twenty years. The circumstances of their extermination in the Connecticut are well known, and the same story, with names and dates changed, serves equally well for other rivers.

"In 1798 a corporation, known as the 'Upper Locks and Canal Company,' built a dam sixteen feet high at Millers River, 100 miles from the mouth of the Connecticut. For two or three years fish were seen in great abundance below the dam, and for perhaps ten years they continued to appear, vainly striving to reach their spawning grounds; but soon the work of the extermination was complete. When, in 1872, a solitary salmon made its appearance, the Saybrook fishermen did not know what it was." [1]

The Pacific salmon is rapidly disappearing also. "Naturally the salmon millions of the Pacific streams early attracted the attention and aroused the avarice of men who exploit the products of nature for gain. As usual, the bountiful supply begat prodigality and wastefulness. The streams nearest to San Francisco were the first to be depleted by reckless overfishing. Regarding the conditions that in 1901 prevailed in Alaska, the following notes are of interest: 'The salmon of Alaska, numerous as they have been and in some places still are, are being destroyed at so wholesale a rate that before long the canning industry must cease to be profitable, and the capital put into the canneries must cease to yield any return.'

"The destruction of the salmon comes about through the competition between the various canneries. Their greed is so great that each strives to catch all the fish there are, and

[1] Jordan and Evermann's "American Food and Game Fishes."

all at one time, in order that its rivals may secure as few as possible. . . . Not only are salmon taken by the steamer load, but in addition millions of other food fish are captured, killed, and thrown away. At times, also, it happens that far greater numbers of salmon are caught than can be used before they spoil. . . . In many of the small Alaskan streams the canning companies built dams or barricades to *prevent the fish from ascending to their spawning beds*, and to catch all of them. In some of the small lakes, the fishermen actually haul their seines on the spawning grounds.

"The laws passed by Congress to prevent the destruction of the Alaskan salmon fisheries are 'ineffective, and there is scarcely a pretense of enforcing them.' To-day the question is — will lawless Americans completely destroy an industry which if properly regulated, will yield annually $13,000,000 of good food? Will the salmon millions of the Pacific share the fate of the buffalo millions of the Great Plains? At present *it seems absolutely certain to come to pass*. . . . The time for strong, effective, and far-reaching action for the protection of that most valuable source of cheap food for the millions is *now!*" [1]

Many of the states have passed laws for the protection and conservation of game fishes such as trout and bass. The sportsmen have seen to this; and while it is desirable that these forms of wild life should be preserved and their number increased in all our waters, it is of much greater importance that the fishes which supply food for the millions should not be left to the mercy of such utterly selfish men as those responsible for the rapid depletion of the Atlantic salmon and the rapid decrease of the Pacific salmon.

It is necessary not only that the number of all fish desirable for food should be increased by means of artificial propaga-

[1] Hornaday's "The American Natural History."

tion as indicated in **105**, but also that wise laws governing the catching of fish should be passed and rigidly enforced. The United States government has done and is still doing splendid work in the artificial propagation and distribution of fishes through the agency of the thirty-nine fish-hatching stations of the Bureau of Fisheries, but has done little or nothing in the regulation of the fish industry. This has been left to the initiative of the states. Following are some of the regulations that many of the states have embodied in laws: (1) There must be no obstruction in rivers that would prevent fish from moving freely up and down streams either to spawn or to search for food. If dams are built, runways must also be constructed permitting the free passage of fish. (2) Fish must not be caught at the spawning season, otherwise the future supply is endangered. (3) No methods of fishing should be employed in which immature fish are caught or killed. Such methods are (*a*) exploding dynamite in the water, thus killing all kinds and sizes of fish indiscriminately; (*b*) catching fishes with nets the meshes of which are so small that immature fish are caught as well as mature; (*c*) wholesale and mechanical devices of catching fish such as the fishing wheel, for by this device the fish have no chance for escape. (4) Fishermen must not keep fish even when caught if they are undersized. (5) It should be illegal to destroy any food fish or use it for any purpose other than food.

These laws are enforced by state fish and game wardens provided there is public demand for their enforcement. The necessity for the enforcement of these regulations will be obvious, not only in waters over which the states have jurisdiction, but also in the waters controlled by the United States government.

CHAPTER V

CRAYFISHES AND THEIR RELATIVES

112. A study of the crayfish. — Laboratory study.

A. *Regions.* — The body of the crayfish has two distinct regions. The dorsal surface and sides of the anterior region are covered by a cape, consisting of a single piece of shell-like material. This region is the *cephalothorax* (from Greek — head-thorax). The posterior [1] region is the *abdomen.*
1. Which region is composed of a number of similar segments?
2. Which region has the legs, antennæ (feelers), and eyes attached to it?

B. *Adaptations for walking.*

Place a crayfish in the center of a pan with enough water to cover the animal. If the crayfish does not walk, touch it with the pincers.
1. How many pairs of legs are used in walking?
2. In what directions (forward, backward, or sideways) are you able to get the crayfish to walk?
3. State whether or not the "large claws" are used in walking.
4. Are the walking legs composed of one piece or of several movable parts? Of what advantage is this to the animal?
5. (Optional.) Make a sketch (×2) of one of the legs to show its structure.

[1] The meaning of each of these terms is explained in **6**.

C. *Adaptations for swimming.*

Place an active crayfish in a pan nearly filled with water. Use the following means to get it to swim: make a sudden movement toward it with the forceps or pencil; if this does not succeed, take hold of the animal near the anterior end where you can press the large pincers against the body. Do this quickly and release the animal. This action may cause the crayfish to swim in order to escape. If you cannot get this crayfish to swim, try another.
In what direction does the crayfish swim?
1. State whether or not the legs are used in swimming.
3. Watch the segments of the abdomen and the large appendages at the posterior end to determine their action in swimming.
   a. Describe the direction of the movements of these parts.
   b. Are these movements made slowly or quickly?
4. In what direction will the doubling under of the abdomen tend to send the animal?
5. In what direction will the straightening out of the abdomen tend to send the animal?
6. In what direction, therefore, must the crayfish strike the harder and quicker in order to swim backwards?
7. What difference is there in the shape of the ventral surface and the dorsal surface of the abdomen?
8. Which surface of the abdomen will enable the crayfish to get the better hold upon the water?

9. (Optional.) Straighten out and double up the segments of the abdomen, noting how the segments are connected. Describe now all the adaptations of the abdomen and its appendages for swimming?
10. (Optional.) The first segment of the abdomen (next to the cephalothorax) fits under the cape; the last is unlike the others in shape, being quite flat. Straighten

CRAYFISHES AND THEIR RELATIVES 153

out and double up the parts of the abdomen; of how many segments is it composed?

11. (Optional.) The large appendages (*large swimmerets*) and the last segment of the abdomen taken together are called the *tail fin*. Make a sketch (× 2) of the abdomen and the large swimmerets. Label: first segment, last segment, large swimmerets, tail fin.

D. *Adaptations for breathing.*

To the Teacher.—Prepare some preserved crayfishes in the following manner: Insert the point of the scissors beneath the posterior margin of the cape that covers the cephalothorax and halfway between the middle line of the dorsal surface and the lower margin of the cape; cut forward to the front end of the cape and remove the piece of shell.

1. Immerse in water a crayfish prepared as directed above. Examine and describe the structures that you find above the legs on the side where the cape has been partially removed. These structures are the special breathing organs of the crayfish. They are known as *gills*.
2. Push the gills to one side and find the soft body wall. Higher up find the line of attachment between the shell and the body wall. You will see that the gills are not inside the body, but in a space between the body of the animal and its shell. This space is called the gill chamber.
    a. In what region of the crayfish are the gill chambers found?
    b. What forms the outer wall of each gill chamber? What forms the inner wall?
    c. Lift up the cape on the opposite side of the animal; state where it is free from the body wall.
3. Examine the gills on a leg that has been removed from the thorax and floated on water and note

that it is largely composed of numerous slender divisions, called the *gill filaments.*

Make a sketch of the leg (× 2) with the gills attached and label gill filaments.

4. The gills are furnished with numerous minute thin-walled blood vessels and the blood in them is separated from the water only by a thin membrane. The blood flows into the gills from all parts of the body by one set of blood vessels and leaves the gills by another. Bearing in mind that breathing is essentially the same in animals as in plants (**P. B., 82**), —

   a. What gas will the blood bring from the body to be given off in the gills in the process of breathing?

   b. What gas is taken up by the blood in the gills to be carried around the body?

   c. How are the gill filaments (as stated above) fitted by structure to permit this interchange of gases?

   d. How are the delicate gill filaments protected from injury?

5. If the same water remained on the gills for some time, what changes in the relative amounts of oxygen and carbon dioxid in the water would occur? Why, then, is it necessary that a current of water should pass over the gills?

6. *Do currents of water pass through the gill chamber?* — Demonstration.

   Inject some harmless coloring matter, such as powdered carmine in water, into the posterior end of the gill chamber. Place the crayfish again in water.

   a. State what was done in this experiment.

   b. Give your observations and conclusion.

   c. What will the incoming current of water bring to the gill filaments?

   d. What will the current of water carry away from the gill filaments?

CRAYFISHES AND THEIR RELATIVES

7. *How the crayfish causes a current of water to pass through the gill chambers.*

   To the Teacher. — Prepare several living crayfish so that the action of the gill bailer may be seen. To do this carefully cut off a small part of the anterior portion of the shell just over the gill scoop.

   Watch the movements of the small blade-like body in the front of the gill chamber. This body is the *gill bailer*, or *gill scoop*.
   a. Describe the movements of the gill scoop, or gill bailer.
   b. When it moves upward and forward, what effect will the gill bailer have on the water in front of it and in the gill chamber?
   c. Where can water enter the gill chamber? (See *D*, 2, *c*.)

8. (Optional.) The gill bailer is a part of one of the crayfish's mouth parts, known as the *second maxilla*. Examine a second maxilla that has been removed from the head thorax of a preserved crayfish. Place it in a watch glass half filled with water and make out the following parts: —
   a. A part shaped something like a bird's wing, composed of several pieces.
   b. The gill bailer that you have already seen.
   c. The part where the second maxilla was torn from the body, clearly shown by the shreds of muscle.
   When you have made out these parts, make a sketch of the second maxilla ($\times 4$), and label: winglike part, gill bailer, shreds of muscle.

9. How does the shape of the gill bailer fit it for the work it does?

E. (Optional.) *Adaptations for food getting.*
   1. Place an earthworm, a piece of beef, or a piece of clam near a crayfish, and describe the way in which he gets the food to his mouth.

2. Of what use may the mouth parts (easily seen in a living crayfish) be in getting food into the mouth?
3. Push the outer mouth parts of a living crayfish to one side with the forceps and find a pair of hard jaws, *mandibles*. Pry them open a little.
   a. Do they work from side to side or up and down?
   b. Describe the cutting edges of the mandibles.
   c. Of what use would these jaws be in preparing food for swallowing?

F (Optional.) *Adaptations for protection.*
1. Describe the outer covering of the animal? Of what use is this to the animal?
2. Locate the softer parts of the crayfish's armor? How are these protected by their position?
3. Gently touch the eye of a living crayfish.
   a. Describe the movements of the eye. How might these movements be advantageous to the animal?
   b. Of what advantage may it be to the crayfish to have its eyes on stalks instead of on the surface of the head?
   c. Make a sketch (× 4) of one of the eyes on its stalk. Label: fleshy stalk, eye.
4. Of what use may the large pincers be in addition to helping in securing food? Sketch (× 1) one of the large pincers complete.

G. (Optional.) *Additional drawings.*
1. Make out the parts of one of the large antennæ. Notice the broad finlike part at the base of the antenna, then two segments, and a long lash that arises from the second segment. Sketch (×2) a large antenna. Label.
2. Make a sketch (× 1) of dorsal view of the crayfish. Label the regions, and all the appendages.

113. Habits of crayfishes. — Crayfish are found commonly throughout the United States in rivers and their tributaries

CRAYFISHES AND THEIR RELATIVES 157

where limestone is found, since lime is needed in making their hard outer covering. During the day they hide under stones, in the crevices of rocks, in the mud, and sometimes in specially constructed burrows along the banks. Since the animal backs into these hiding places, its big claws are ready for business if an enemy attacks it.

Then, too, the colors of crayfishes aid somewhat in protecting them since these colors are usually similar to the color of the bottoms of the streams in which they live. Lastly, the wide range of vision, which the stalked eyes afford must serve to warn the animal of the approach of danger. Nevertheless they do not always escape since crayfish are often captured by certain birds and fishes. In fact, crayfishes are often used by man as a bait for catching fishes.

114. Food, food getting, and digestion. — At night crayfishes crawl about in search of food, concerning which they are not at all fastidious, since dead fish and other dead animals seem to be fully as acceptable as when alive. In fact, they are natural scavengers. Crayfish seize their food with their large claws and with the aid of the small pincers on the front walking legs and with the mouth parts, especially the mandibles, reduce the food to pieces small enough to be eaten. We have seen in plants (P. B., 63, 70, 71) that digestion many take place in any living cell where food is stored or manufactured. Hence, plants have no special part devoted to digestion. In crayfishes, however, it is quite different, since a part of the body, known as a *digestive system*, is devoted to preparing the food for absorption and use. This digestive system consists of a food tube known as the *alimentary canal* and certain masses of cells known as *digestive glands*.

After the food is digested, it can pass into the blood by

osmosis and be carried to the cells [1] of the body. When the digested food reaches the cells, it may be used by the protoplasm either in making more living matter or, as we shall now see, for the release of energy.

115. Respiration and the production of energy. — In our laboratory study we watched the movements of the gill bailer and saw that it caused a current of water to enter the posterior end of the gill chamber and flow over the gills, thus bringing oxygen to the filaments (Fig. 111). The thin-walled blood vessels in the filaments absorb the oxygen, and the blood then passes on into other blood vessels, which carry it back to the heart, whence it is forced all over the body of the crayfish, and so the oxygen in the blood passes into the cells as does the food. Now what becomes of the oxygen?

Fig. 111. — Gills of a crayfish.

As in plants (P. B., 80), the oxygen unites with elements in the foods and protoplasm of the cells and produces oxidation and liberation of energy, which gives the crayfish the power to contract its muscles and so push against the water with its abdomen and tail fin, thus propelling the animal backward, or to open its nippers and shut them and so secure food. In fact, all the work that the crayfish performs is made possible through the burning of its foods or protoplasm by the oxygen.

[1] We have shown in plant biology (41) that plants consist of cells which are largely composed of living matter known as protoplasm. This is also true of animals (126).

CRAYFISHES AND THEIR RELATIVES

Since the proteins, fats, carbohydrates, and protoplasm all contain carbon, when these are oxidized, carbon dioxid (CO_2) will be formed as one of the waste substances. All these waste substances will pass out of the cells into the blood, which finally conveys them to the filaments of the gills. Here the waste matters pass out into the water, which, as we have seen, is then forced out of the front end of the gill chamber.

116. Life history. — As in the seed-producing plants, crayfishes are reproduced by means of special cells known as egg-cells which in crayfishes are formed in the body of the female in organs known as ovaries. Before they can develop, however, these egg-cells, as in the seed plants, must be fertilized by sperm-cells, produced in spermaries of the male crayfish. After extrusion the fertilized eggs are attached by a sticky substance to small appendages, known as swimmerets, on the ventral surface of the abdomen of the female (Fig. 112). Here the fertilized egg-cell develops into a many-celled embryo, and finally a tiny crayfish is hatched. At first the young crayfishes are held to the swimmerets by threads; later they cling by means of their pincers, and after some days become independent. At intervals in both young and old crayfishes, the hard outer covering of the body is shed. This shedding of the skin is called *molting*. But for this process it would be impossible for the young to grow.

While the young crayfishes are attached to the parent they are of course protected by their position, and the female looks after them by looking out for herself. The food for the developing embryo is stored in the egg. After hatching, the young must care for themselves, and after they become independent they receive no protection at all. There is, therefore, in the case of crayfishes nothing like the parental care of higher animals.

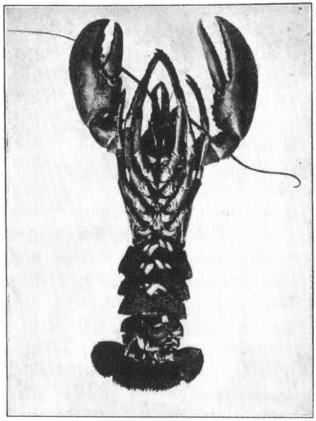

Fig. 112. — Female lobster with eggs beneath abdomen. (Herrick's "American Lobster" — United States Fish Commission.)

117. Relatives of the crayfish. — One of the relatives of the crayfish is the lobster (Fig. 112), which is a salt water animal found along the north Atlantic coast. Like the crayfish, its body consists of a cephalothorax and a clearly segmented abdomen. The lobster also has two pairs of antennæ, a pair of stalked eyes, a number of pairs of mouth parts, a pair of big claws, four pairs of walking legs, to the bases of which gills are attached, and a pair of swimmerets

FIG. 113. — The crab.

on each of the segments of the abdomen except the last. In general, lobsters are very much larger than crayfishes, one of the largest known specimens weighing over twenty-three pounds.

Less like the crayfish in appearance are the crabs, yet a careful examination shows that these animals have practically all of the characteristics mentioned in the preceding paragraph. The cephalothorax of crabs, however, is usually wider than it is long (Fig. 113), and the abdomen is much reduced and is commonly folded in a groove beneath the cephalothorax. Few of the crabs are able to swim; usually they crawl sideways by the help of their four pairs of walking legs.

FIG. 114. — The hermit crab in an empty snail shell.

"A curious modification of habit is shown in the hermit crab (Fig. 114), which in early life backs into an empty snail shell which aids in protecting it from its enemies. The abdomen, thus covered, becomes soft and flabby. As growth proceeds the necessity arises for a larger shell, and the crab goes 'house-hunting' among the empty shells along the shore, or it may forcibly extract the snail or other hermit from the home which strikes its fancy." — JORDAN and HEATH, "Animal Forms."

M

Among the relatives of the crayfish that live in damp places on land are the pill bug and the sow bug (Fig. 115) which are often

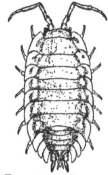

FIG. 115.—The sow bug.

found beneath water-soaked wood. All the animals we have described in this chapter belong to the class *Crustacea,* so-called from the hard outer shell which invests them.

118. Economic importance of the Crustacea.—Crayfishes in Europe, particularly in France, are highly esteemed as food, and special efforts are made to increase their number. In this country, however, they have, as yet, been used but little as food. Their principal use is for bait in catching certain kinds of fish.

The lobster is to us what the crayfish is to Europeans. While they are not abundant enough to be considered a very important source of food, still the fishermen in 1901 received $1,400,000 for the lobsters

FIG. 116.—The shrimp.

caught. They are considered rather as a delicacy, since they are too expensive for general use, principally on account of their scarcity. For a number of years the United States government has been making efforts to increase the number of lobsters by artificial propagation. Some states have passed laws forbidding the catching of immature lobsters and lobsters with eggs attached.

Other crustacea that are used for food are prawns, shrimps (Fig. 116), and certain kinds of crabs. Nearly all the crustacea eat dead animal food; consequently they are useful in keeping the water free from dead material.

CHAPTER VI

PARAMECIUM AND ITS RELATIVES

119. Study of the paramecium. — Laboratory study.

Note to Teacher. — To secure paramecium material, add some chopped hay to a large jar of water several weeks before the animals are needed. The paramecia develop more rapidly and are of larger size if the water is secured from a stagnant pool. The hay infusion furnishes food for bacteria upon which the single-celled animals feed. To obtain the paramecia, transfer to a glass slide with a pipette a drop from near the surface of the water.

A. *General appearance of paramecium.*
1. Place a drop of water containing many paramecia or other similar forms on a glass slide (with concave depression if possible). Examine with a magnifier.
 Describe the appearance of the tiny bodies that you see moving about.
2. Now examine the drop of water with the low power of the compound microscope. Do not allow the water to evaporate entirely, but keep adding a little from time to time.
 a. Do the paramecia swim slowly or rapidly?
 b. Is the more pointed end of the animal usually foremost in swimming or the rounded end?

B. *Structure of paramecium.*
 Secure a stained and mounted specimen of a paramecium, or add a drop of iodine solution to the water containing the living animals, and

PARAMECIUM AND ITS RELATIVES

place a cover glass on top. Examine first with the low power of the microscope and then with the high power. Make a sketch two or three inches long to show the following: —
1. The general shape of one of the paramecia.
2. A fringe of slender hairlike projections around the outer surface. They are called *cilia* (singular *cilium*, from Latin, meaning a hair). The cilia are projections of the protoplasm of the cell. They project from the upper and lower surface also, but they cannot be seen readily.
3. A more deeply stained portion of the protoplasm near the center, the *nucleus* (Fig. 118). The rest of the cell is the *cell body*.
4. Particles of matter, *food particles* scattered through the body of the cell.
5. Label: cilia, nucleus, food particles, cell body.

C. *Food getting*.

To the drop of water containing the living paramecia add a little finely powdered carmine, and on the drop place a cover glass.
1. Tell what was done.
2. Throw all the light you can on the paramecia by means of the mirror and use the larger openings in the diaphragm. What evidence have you that the paramecia are feeding on the carmine? Sometimes it is necessary to leave the paramecia for twenty-four hours before they feed.
3. Watch the paramecia swimming through the particles of carmine. What evidence have you that the cilia are in motion?
4. The paramecium has a *furrow* on one side of its body, and from the furrow a tubular passage or *gullet* leads into the protoplasm. Both the furrow and the gullet are lined with cilia.
   a. If you are able to see either the furrow or the gullet, describe them.
   b. In what direction must the cilia in the furrow and

the gullet strike the swifter and with the more force to bring food particles into the gullet?

D. *Locomotion.* (Optional demonstration.)

Examine with a high power a paramecium that is comparatively quiet. Focus carefully and look for the cilia.
1. Describe the cilia and their movements.
2. When the paramecium strikes against the water in one direction, in what direction would its body tend to move?
3. Must the paramecium strike harder toward the blunt end or toward the pointed end when it swims with the blunt end foremost?

E. *Excretion of liquid waste.* (Optional demonstration.)

Look at a paramecium or vorticella with both the low and high power and search for clear circular spots. Watch to see if any

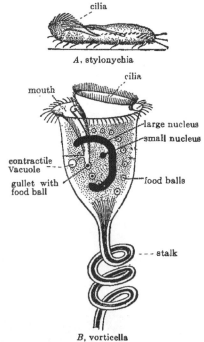

Fig. 117. — Protozoa with cilia.

of these contract. If they do, they are *contractile vacuoles.* There are two in paramecium and one

PARAMECIUM AND ITS RELATIVES 167

in vorticella (Fig. 117). The liquid waste flows from the protoplasm into these spaces, the protoplasm then pushes together and forces the waste out of the body.

1. Describe the position, appearance, and action of the contractile vacuoles.
2. State in your own words the use of the contractile vacuoles.
3. Sketch the contractile vacuoles in your drawing of the paramecium and label.

F *Reproduction of paramecium.* (Optional demonstration.)

All the time while you are studying the paramecium be on the lookout for forms that are dividing. If you do not see any, examine mounted slides that show the paramecium dividing. Make a sketch three inches long of a paramecium dividing, to show how it reproduces.

120. External structure and locomotion. — In form a paramecium resembles somewhat the shape of a slipper, hence it is sometimes called the "slipper-animal" (Fig. 118).

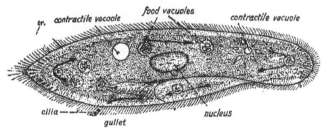

FIG. 118. — The paramecium. (Dahlgren.)

Extending from all parts of its outer surface are many tiny projections of protoplasm that look like colorless hairs; these are known as *cilia* (singular *cilium*). In locomotion the animal usually moves with the blunt end (*i.e.* heel of the slipper) in front, the paramecium being propelled by the strong backward strokes of the cilia and a slower recovery. When it runs into an obstacle, the cilia are reversed in action and thus the animal is enabled to move with the opposite end (toe of slipper) in front. Most animals that swim (*e.g.* fishes and frogs) have broad and flat appendages which are comparatively large. In paramecium, on the other hand, the organs of locomotion (cilia), while slender, are so numerous that they perhaps accomplish the same results as the broad swimming appendages of the frogs and fishes.

121. Food, food getting, and digestion. — Paramecia feed upon one-celled plants and animals. On one side of a paramecium is a furrow or *groove*, which is lined with cilia. At the lower end of the groove is an opening, the *mouth*, which leads into a short, tubular *gullet*. The rapid motion of the cilia in the groove draws the food toward the mouth opening and other cilia lining the gullet push down the food particles. Small collections of these food particles are made at the lower end of the gullet, and these masses, *food balls*, are circulated within the cell by the streaming movement of the protoplasm. Although the paramecium is a single cell, it has certain parts specially developed for securing food, just as the higher animals have special organs for this function.

As the food balls circulate through the protoplasm, they are gradually digested, and the food materials thus liquefied are used as in plants and other animals for the production of more protoplasm or for the release of the energy needed for locomotion and for food getting. The indigestible parts of food are forced out through the side of the body.

122. Respiration and the liberation of energy. — The paramecium is surrounded by water that contains oxygen and this passes into the protoplasm through the thin membrane surrounding the animal. When the oxygen combines with the chemical elements found in foods and protoplasm, oxidation is carried on, energy is released, and waste substances are formed which are given off in the process of excretion.

123. Excretion of wastes. — At either end of the animal is a clear space which is sometimes circular and at other times star-shaped. These are the *contractile vacuoles*. The wastes formed by oxidation (*e.g.* carbon dioxid and water) collect to form the vacuoles. The protoplasm presses upon the waste materials and periodically squeezes them out of the animal. When this occurs, the contractile vacuole disappears.

124. Reproduction and life history. — In the interior of a paramecium are two nuclei known as the large nucleus and the small nucleus, both of which show readily when the animal is stained with iodine or with other chemicals. When the animal reproduces, both the large and small nuclei divide in halves (Fig. 119), a new mouth and gullet are formed, and two new contractile vacuoles appear. The cell body then divides transversely, the cells separate from each other, and thus from a single individual, two new paramecia are formed. If conditions are favorable, both animals grow and may in turn reproduce at the end of twenty-four hours. "It has been estimated that one paramecium

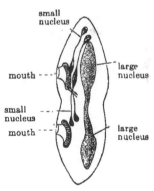

FIG. 119. — A paramecium dividing.

may be responsible for the production of 268,000,000 offspring in one month."

125. Study of amœba (plural, amœbæ or amœbas). — (Optional laboratory study.)

A. *Structure of amœba.*

Examine a living amœba or a stained specimen on a prepared slide. Use a low power of the compound microscope at first, and then as high a power as may be necessary. Make a sketch about three inches long to show the following : —

1. An outline to show the shape of the animal, including any projections of the protoplasm, which are called *pseudopods* (Greek *pseudo* = false + *pod* = foot; hence, the name *false foot*).
2. The main mass of the amœba, clear and jellylike in a living amœba, slightly stained in a mounted specimen, which is called the *cell body.*
3. A slightly denser part of the protoplasm in the living form or stained much darker in the preserved animal, the *nucleus.*
4. Particles of food or one-celled plants scattered through the cell body.
5. Label: false feet or pseudopods, nucleus, cell body, food particles.
6. If time allows, draw several different forms assumed by the specimen.

B. *Locomotion.*

In a living amœba watch with the high power of the microscope the creeping movements, and the projections of the pseudopods.

1. Are the movements slow or rapid?
2. In your own words give a description of the locomotion of the amœba.

PARAMECIUM AND ITS RELATIVES

C. Excretion of liquid waste.

Look for a clear, roundish spot in the amœba which at intervals disappears. This is the *contractile vacuole.* The liquid waste flows into this space and then the protoplasm pushes together and forces the waste out of the body.

1. Describe in your own words the appearance and action of the contractile vacuole.
2. Sketch the contractile vacuole in your drawing of the amœba and label.

126. A comparison of paramecium and amœba. — Both amœba and paramecium are animals so small that they can barely be seen with the naked eye. Both live in water, both are one-celled animals, and both carry on the same functions, but in a somewhat different manner. While the paramecium maintains a more or less fixed form, the amœba is capable of assuming almost any shape (Fig. 120). This it does by causing portions of its substance to flow out in many directions. These projections are known as *pseudopods* which mean *false feet.* By pushing out these pseudopods in front and pulling up its protoplasm from behind, the amœba slowly flows from one part of the slide to another.

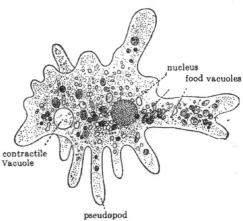

FIG. 120. — The amœba.

Unlike the paramecium an amœba has no definite part of the body through which it takes in food. When the animal is feeding,

172 ANIMAL BIOLOGY

it slowly flows about the one-celled plant or animal and finally ingulfs it. The processes of digestion, assimilation, respiration, excretion, and reproduction (Fig. 121) are much the same in amœba as in paramecium. Both these animals belong to a group of animals known as the *Protozoa* (Greek *protos* = first or simplest + *zoön* = animal).

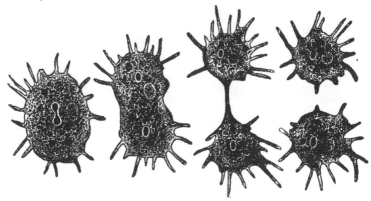

Fig. 121. — An amœba dividing.

127. To show that the higher animals are composed of many cells. — Laboratory study.

Frogs are continually shedding parts of their epidermis, and pieces of this thin membrane are likely to be seen clinging to a frog in an aquarium or floating in the water. Secure a piece of this membrane, spread it on a slide, add a drop of water and a cover glass, and examine with the low power of the microscope.

1. Describe the form and color of each cell.
2. In each cell notice a body, usually near the center and slightly more dense than the rest of the cell. This is the *cell nucleus*. (If the nucleus does not show clearly, add a drop of iodine to the membrane.) The rest of the cell is the *cell body*.
    a. Name, now, two parts of a cell of the frog's epidermis.

b. State the form and position of the cell nucleus.
3. Make a drawing of three of the cells described above, each cell to be represented about an inch in diameter. Label cell body and cell nucleus.
4. (Optional.) Demonstrate by the use of prepared slides, pictures, or charts that the blood, intestine, and other organs of the body of a frog or other higher animal are composed of cells. Make a drawing of a single cell in each case.

128. A comparison of Protozoa and the higher animals. — Our study thus far has shown that all animals, including the Protozoa, perform the necessary functions of locomotion, food getting, assimilation, respiration, and reproduction. The adaptations for performing these functions, however, are very diverse.

All animals except the Protozoa consist of many cells and the various functions of the higher animals are performed by groups of cells known as *organs*. For example, certain combinations of cells carry on locomotion, others digestion, while still others are set apart for breathing. All these functions are performed in a Protozoan by a single cell.

129. Economic importance of Protozoa. — Most of the Protozoa serve as food for other animals that live in the water and these in turn are fed upon by fish, which are eaten by man. Thus the one-celled plants and animals are found to be an important food-basis for human beings.

Some of the Protozoa that live in the sea secrete tiny shells (Fig. 122), and when the animals die the shells drop to the bottom. As a result of heat, pressure, and other causes, this bottom ooze is gradually solidified to form chalky rocks, and in the upheavals that have taken place in ages past these rocks have been forced above sea level. The chalk cliffs of Dover, England, were doubtless formed in this way.

While most of the Protozoa are harmless, there are a few forms that have become parasitic in human beings. We have already discussed the single-celled animal that causes malaria and that is carried from one individual to another by the Anopheles mosquito (39). This parasite resembles an amœba in form. Another form of Protozoan causes the terrible disease known as the sleeping sickness of tropical Africa. Many biologists believe that yellow fever (41) is caused by a protozoan that is transmitted by the Stegomyia mosquito.

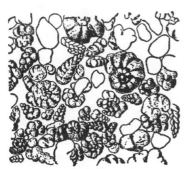

Fig. 122. — The shells of one-celled animals as they are found in chalk. (Scott, Geological Survey of Iowa.)

CHAPTER VII

ADDITIONAL ANIMAL STUDIES

A. *Porifera* (*sponges*)

130. Sponges. — The sponges are animals more complex in structure than the Protozoa, for they are composed of many cells; nevertheless, they are comparatively simple in structure since they have no digestive, circulatory, respiratory, or nervous system, and therefore each cell has to carry on practically all the necessary nutritive functions.

Sponges differ largely in the kind of skeletons that they possess. In the common bath sponge (Fig. 123) this is composed of a tough, horny material. When sponges are ready for market, only the horny skeleton remains, the living cells having been killed and removed. The sponge skeleton shows a large number of pores in the outer surface, and for this reason the name *Porifera* (Latin = pore-bearing) is given to this group of animals. The pores lead into canals that run through the body, finally connecting with one or more larger central cavities that lead outward, usually at the top. In certain parts of these canals there are cells with cilia; their action causes water to rush into the canals through the pores, bringing food and oxygen to all the cells of which the sponge is

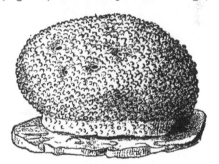

Fig. 123. — Bath sponge.

composed. The wastes are forced out through the larger canals referred to above. Like the bath sponge, all other Porifera are stationary in their mature form.

B. Cœlenterata

131. Hydra. — A study of a fresh water cœlenterate known as hydra will give one a fair idea of the structure and adaptations of this group of animals. Hydra is a small animal found in fresh water attached to water plants, and sometimes to surfaces of stones or

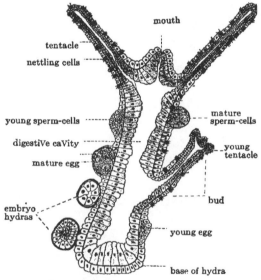

FIG. 124. — Longitudinal section of a hydra. (Hegner.)

other objects on the bottom. At the upper end of the tiny cylindrical *column* are threadlike bodies known as *tentacles* (Fig. 125, *1*). If the animal is touched with a needle or pencil, it contracts its body and tentacles so much that it can scarcely be seen. But in a short time it expands again.

If the hydra happens to be hungry and some small form of animal comes in contact with the waving tentacles, the hydra ejects microscopic threads from certain cells (*nettling cells*) in the tentacles. The animal thus attacked is benumbed, and the hydra then uses the tentacles to push its prey into a mouth opening in the center of the circular row of tentacles. The food is drawn into the inside of the column, which is simply a hollow tube (Fig. 124). Here certain cells secrete digestive ferments which dissolve the foods that the animal has eaten, and the indigestible matter is ejected from the mouth. The digested food is then absorbed by the cells lining the cavity. Since the animal is bathed outside and inside by water containing oxygen, the cells are able to absorb oxygen from the water and to give off carbon dioxide to the water. Hence no breathing organs are needed.

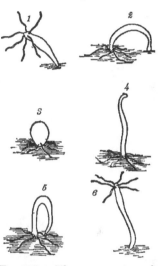

Fig. 125. — The movements made by hydra in locomotion. (Jennings.)

It is evident that the tentacles with the nettling cells also serve to protect the hydra from too great familiarity on the part of visitors that might otherwise use it for food. When the hydra moves from one place to another, it bends over until the ends of the tentacles touch the surface on which it rests. The tentacles then adhere to this surface, the bottom of the column lets go, and the animal turns a somersault (Fig. 125) and lands on the lower part of the column; the process may then be again repeated.

Like the higher animals the hydra reproduces by means of eggs and sperms. But it also has another interesting way of producing new individuals. On the surface of the column one frequently sees little bunches. These are called *buds* (Fig. 124). They keep on growing outward till at last little tentacles and a mouth opening are formed at the tip of each. It is now evident that we are looking at a very tiny hydra. Finally the new individuals separate from the column and begin an independent life. This method of reproduction is known as *budding*.

A, organ-pipe coral   *B*, precious coral   *C*, sea-feather

Fig. 126. — Different forms of coral.

132. Suggestions for the study of hydra. — Laboratory study. Pupils should be supplied with living hydra if possible. The column and tentacles should be observed by the aid of a magnifier, described and drawn. The animal should be touched and the action of the column and tentacles noted and described. If the hydra moves from place to place, the method of locomotion should also be described.

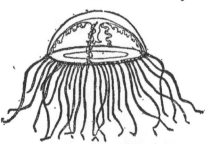

Fig. 127. — Jellyfish. (Hargitt.)

133. Relatives of hydra. — Among the relatives of hydra are the corals (Fig. 126), sea-anemones, and jellyfish (Fig. 127). One form of coral, the red coral, is of considerable economic importance. In all the corals the column secretes a mineral substance within which the animal can withdraw when danger threatens. In the case of the red coral this material is horny. It is used for decoration, and some communities on the Mediterranean are devoted largely to the gathering of this coral, and to making it into various forms of jewelry.

C. Annelida

134. Earthworm. — The most common representative of the annelida is the earthworm (Fig. 128). The general form of this animal is long and cylindrical. If one places an earthworm on the

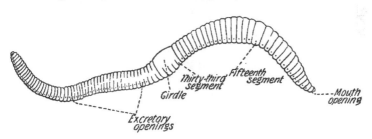

FIG. 128. — The earthworm. (Sedgwick and Wilson.)

ground, it will start to crawl away or bore into the soil. Observe that the end that is foremost is tapering. This is the *anterior end*. The opposite or *posterior end* is broader and considerably flattened. The part of the body on which the worm crawls is the ventral surface, which is somewhat flattened, while the dorsal surface is rounded. The whole body is composed of rings or *segments*. About one third of the distance from the anterior end of the worm several of the segments are usually somewhat enlarged and form the *girdle*.

At the anterior end toward the ventral surface, there is a small opening. This is the mouth, and through it the earthworm sucks in its food which consists not only of dirt, but of leaves of various kinds. Overhanging the mouth is a tiny projection, the *lip*. The animal has no special breathing organs. The skin, however, is permeated with capillaries, and thus serves as a breathing organ.

Locomotion is brought about by alternately lengthening and then shortening one portion of the body after another. On the ventral region of the body are rows of bristles which aid in locomotion. The bristles project backward when the worm is moving forward, and so keep the animal from slipping backward when it lengthens itself. The bristles also serve to hold the animal in its burrow.

Earthworms are of considerable value in the soil. They burrow through the earth by swallowing the dirt which is mixed with vegetable matter; both are then acted upon by digestive juices in the alimentary canal. The refuse of the food, which is not available for use in the body, is ejected from the posterior end of the intestine. The little piles of dirt that are sometimes so common on a lawn are the "castings" of earthworms. It has been found that soil worked over by these animals is in better condition for the growth of plants. Then, too, the deeper soil that has not been used by plants is brought to the surface and mingled with the dirt recently used. Darwin [1] estimated that in England earthworms annually bring to the top of the ground eighteen tons of soil per acre.

135. Suggestions for the study of the earthworm. — Laboratory study.

This study should be made upon living worms. The pupil should first note and describe the general shape and segmentation of the animal, the differences between the anterior and posterior ends, the dorsal and ventral surfaces, and the characteristic appearance of the girdle. An earthworm should be placed on a moist surface such as soil or wet paper, and the locomotion of the animal observed and described. A large specimen should be pulled, anterior end first, between the fingers, and the action of the bristles noted and their situation and appearance studied with the help of a magnifier. Touch the earthworm on various parts of the body, and determine, if possible, which portions are the most sensitive. Look on the dorsal and ventral surfaces for blood vessels, and watch the pulsations of the blood in these vessels; describe the location of these blood vessels and state the direction in which blood flows in each of them.

[1] Darwin's "Vegetable Mold and Earthworms."

ADDITIONAL ANIMAL STUDIES

136. Relatives of the earthworm. — Two forms of animals that formerly were classed with the earthworm under the head of "worms" are the tapeworm (Fig. 129) and trichina. The tape worm is sometimes present in beef and trichina (Fig. 130) in pork. Meats, therefore, should be well cooked to kill all such parasites. The trichina, if it gets into the human system, causes great suffering. When a tapeworm becomes attached to the human intestine by the suckers and hooks on its anterior end, it is difficult to dislodge.

D. Mollusca

137. Fresh water mussel. — The fresh water mussels are mollusks that are sometimes called clams. They are often quite abundant on the bottom of creeks, rivers, ponds, or lakes. Usually they are partly covered with sand or mud, sometimes even more than is shown in Figure 131. It will be seen at once that the mussel is inclosed by a shell. This consists of two parts called *valves;* hence these animals, as well as salt water mussels, clams, and oysters are called *bivalves* (Latin *bis* = two + valve). The two valves are held together along one margin by a tough material that serves as a hinge. On each valve near the hinge, a prominence, known as the *beak* or *umbo*, may be readily seen. Around

A, head of tapeworm

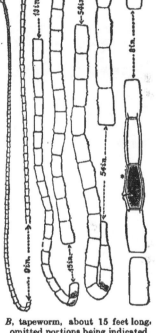

B, tapeworm, about 15 feet long, omitted portions being indicated

FIG. 129. — The tapeworm. (Shipley and MacBride.)

the umbo, in ever widening concentric rings, are the lines of growth of the animal, which indicate younger stages in its development.

Let us now pull up a mussel and lay it on a sandy bottom. In a few moments the shell will open somewhat and from one end will project a pinkish body, which may finally extend some distance. This organ is the *foot*. If we watch long enough, we may see the mussel use the foot to push itself over the surface of the sand or it may burrow into the sand, and finally come to occupy a position like that in which we found it.

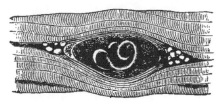

FIG. 130.—Trichina in Muscle. (Leuckart.)

Now if one is patient, and the animal feels at home, it will be possible to see the method of eating and breathing. At the end opposite the foot there may slightly project from the shell a fringed and somewhat tubular-shaped structure. Let us place a little finely powdered carmine in the water above the opening. As the carmine slowly sinks and comes opposite the tube, the particles will suddenly be drawn into the tube. This shows that water is being sucked into the tube, and it brings with it oxygen and any food that may be near, such as microscopic plants and animals.

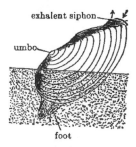

FIG. 131.—Mussel burrowing in sand.

To learn any more about the feeding and breathing of the mussel it will be necessary to open the shell. Let us take another mollusk and pry open the valves. We shall soon find that this is not easy to do. The reason will be evident after studying Figure 132.

The valves are held together by strong muscles. So we pry the valves open a little with a heavy knife and then slip another sharp knife in close to the valve, where we meet an obstruction toward one end. When we have cut this, the valve opens at that end. After cutting the muscle at the other end, we can readily separate the valves. All over the surface of the animal, except where the two muscles were attached to the shell, is a thin covering called the *mantle*. By raising the body of the mussel from the valve it will be evident that there is a similar structure on the other side.

Now, if we fold back the mantle, it will be possible to follow the course of the food and water. The first thing that strikes our at-

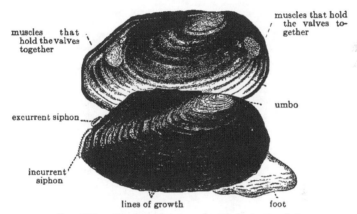

Fig. 132. — Fresh water mussel with foot extended.

tention is the contracted foot, and above this is a soft mass called the *abdomen*. In the abdomen are found the *digestive organs*. On each side of the abdomen are two broad, thin flaps, the *gills*, by which the animal breathes. Between the foot and the end that was buried in the sand are found, on either side of the body, two small flaps or *palps*, and between them lies the *mouth opening*. To this mouth the food that has been swept into the tube is brought by the waving of thousands of cilia that are found on the surface cells of the gills and palps.

Let us now return to the study of the mussel partly covered by the sand. The hinge is on the dorsal region of the body, the free edges of the valves on the ventral, while the mouth and foot are at the anterior end. Hence, the animal in its natural position "stands on its head," or at least where its head ought to be. From the posterior end projects the tubular structure to which reference has been made.

Let us again drop some powdered carmine closer to the animal, and watch the particles when they reach a point just above the tube where we saw the particles enter. We shall now see the carmine carried away from the animal instead of into it. A closer examination reveals the fact that the tubular structure has a second opening above the first. Both of these tubes are called *siphons*, the lower being the *incurrent siphon*, and the upper the *excurrent siphon*. The stream of water forced out of the excurrent siphon carries with it the carbon dioxid and other wastes of the body.

138. Suggestions for study of the mussel. — It is desirable to have students see the mussel in its natural home. They should tell where they found the animals and the positions in which they were seen. It would then be well for the pupil to study in the laboratory the shell, making out the points of structure described above. A drawing of a side view of the mussel should be made and labeled as follows: valve, umbo, hinge, lines of growth, anterior region, posterior region, dorsal edge, ventral edge. It is also desirable that a drawing of the animal in the sand or mud be made and the incurrent and excurrent siphon openings be labeled.

The pupil might well follow the account as given above, verifying the statements and experiments, and making drawings of the mussel with the shell open and all the animal lying in one valve. Label: mantle, muscles that close shell, incurrent siphon, excurrent siphon. Also a drawing should be made of the mussel with the mantle removed. Label: foot, abdomen, palps, mouth, gills. Write an account of how the mussel moves or burrows, how it feeds and breathes.

139. Relatives of the mussel. — Some of the relatives of the mussel are the clams, oysters, salt water mussels, snails (Fig. 133), and slugs. While the fresh water mussels are not much used for food, they are important economically on account of the pearly matter that is found on the inside of their shells. This is used in making buttons and other articles. In fact, there is a considerable industry in this line along the Mississippi River.

Oysters are important as an article of food. The oyster fishermen receive annually from twenty to thirty millions of dollars from these mollusks collected from the oyster beds along the Atlantic Coast. A certain kind of mollusk, known as the *pearl* oyster, secretes within its shell the pearls of commerce. These are formed of a material similar to that found on the inner layers of the fresh water mussel.

FIG. 133. — The snail.

E. Reptiles

140. The turtle. — The body of a turtle may be divided into four regions; namely, head, neck, trunk, and tail. The larger part of a turtle, the trunk, is covered by a shell, and to this shell the bony skeleton is firmly united. The two pairs of legs, however, are freely movable, but can be drawn within the shell for protection. The toes of the feet are armed with sharp, curved nails, and the legs are covered with scales. The legs are used for walking and also for swimming. In some turtles the legs become broad and flat and are of but little use except for swimming.

The head, neck, and tail can also be drawn into the shell. Scales cover the neck and part of the head. The jaws of the turtle, often called the beak, possess no teeth. The eyes, protected by the eyelids, the nostrils, and the ear openings, are readily seen.

Turtles reproduce by means of eggs, which are comparatively large. Turtle eggs are often used for food. These animals breathe throughout their entire life by means of lungs.

141. Suggestions for the study of the turtle. — Turtles are easily kept at home or in the laboratory. The pupil should verify the

Fig. 134. — Lizard of the Southwest (commonly known as the "horned toad").

statements given above concerning the turtle, and should then write an account of his observations in his notebook; or a well-labeled drawing will cover most of the ground. The pupil should also observe and describe in his notebook the methods by which the turtle feeds, crawls, swims, and protects its head, legs, and tail.

142. Relatives of the turtle. — Animals related to the turtle are the lizards (Fig. 134), alligators and crocodiles, and snakes (Fig. 135), all of these animals being known as reptiles. None of the reptiles, other than the turtles, possess a shell, but all are covered with scales, and have toes armed with claws, except

Fig. 135. — The rattlesnake.

the snakes which have no appendages. Unlike the turtles the jaws of all other reptiles contain sharp teeth, used in holding their prey, and in the rattlesnake and copperhead some of these teeth are provided with poison glands. None of the other reptiles in the northern part of the United States are in any way dangerous to man. Indeed, many snakes destroy large numbers of rats and mice, while lizards catch large numbers of insects. The hide of the alligator is of considerable value for leather. All reptiles breathe throughout their life by lungs, and most of them reproduce by eggs, which are hatched by the warmth of the sun.

F. Mammals

143. Characteristics of mammals. — In this class of vertebrates are included domesticated animals such as the cow, sheep, horse, camel, dog, and cat. Let us consider the structure of some of these

Fig. 136. — The sperm whale.

animals to see why they should be grouped together. We are familiar enough with the animals named above to know that they all have a head, neck, trunk, and tail and that these regions are covered with hair. A few mammals, *e.g.* the baboons, have no tail, and a few are nearly destitute of hair, like the whales (Fig. 136); but all of them nourish their young on milk produced in certain organs known as mammary glands; hence these animals are called *mammals*.

The organs of the head, namely the ears, eyes with eyelids, and the nostrils, are prominent in all common mammals, but vary in size and shape. The jaws have teeth set in sockets, but the number and kinds of teeth vary greatly. Rats, rabbits, and squirrels, for example, have sharp cutting teeth (*incisors*) and grinding teeth (*molars*). Others, *e.g.* dogs, cats, lions, and tigers, have sharp pointed incisors and molars and in addition long *canine* teeth for tearing their food. In horses, cows, and other herbivorous animals the grinding teeth are especially developed, while canine teeth are either wanting or are relatively small.

All these animals have four legs, but the relative size of the front and hind legs may differ greatly. In a kangaroo, for instance, the

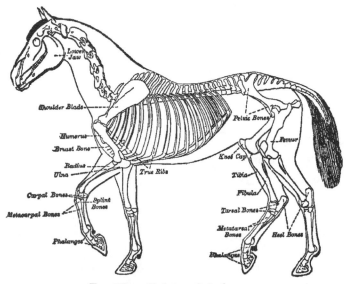

FIG. 137. — Skeleton of the horse.

hind legs are very large, while the front pair are so small as to be practically useless. Then, too, the nails on the toes vary considerably. The fingers and toes of man are protected on a surface by nails. A horse has only one toe on each foot, and the nail for that toe is developed into a *hoof*. Cows and sheep have two toes on each foot similarly protected. On this account these mammals and others like them are called the *hoofed mammals*.

ADDITIONAL ANIMAL STUDIES 189

An examination of the skeleton of a horse (Fig. 137) or of most mammals, shows that the skeleton consists of bones similar to those of man. Thus, for instance, there is the spinal column made up of a series of more or less similar bones, with a skull that may vary a great deal in shape from that of man, but still may consist of similar bones. The shoulder bones and hip bones can be readily distinguished. The bones of the legs are for the most part much alike, but in the foot there is frequently a wide variation, as in the case of the one-toed foot of the horse, the two-toed foot of a cow, the three toes of the tapir, the four of a hippopotamus, and the five of the dog or of man.

144. Suggestions for the study of a mammal.—Follow the general account given above and describe the corresponding structures of a horse, dog, cat, or other mammal. Thus, for instance, name the regions present, and describe the character of the covering of each region. Then describe the situation and parts of the eyes, the situation, size, and shape of the external ears, the location of the nostrils, and so on to the end of the study. Lastly, describe the methods of locomotion of the animal, and its food and feeding habits.

145. Economic importance of mammals.—The mammals include many of our most useful animals as well as those that are very dangerous. Our common beasts of burden, horses and mules in this country, the llama of South America, the elephant and camel of Asia and Africa, are all mammals. This group of animals also furnishes us with an immense amount of material valuable for food or clothing (*e.g.* the cow, deer, sheep, pig, seal). The group of carnivorous mammals contains one of man's most devoted friends and protectors, the dog. To the same order as the dog, however, belong the wolves, lions, tigers, hyenas, and wild cats; all these have canine teeth which they use with deadly effect in tearing their prey. The gnawing mammals (*e.g.* rats and mice) besides being a nuisance, do a great deal of damage. The rat also scatters diseases like cholera and bubonic plague. Some rodents, the beaver, for example, are valuable on account of their fur. The rapacity of man, however, has nearly exterminated these very interesting animals.

G. Classification of Animals

146. Vertebrates and invertebrates. — All animals may be divided into two great groups, known respectively as *vertebrates* and *invertebrates*. To the first group belong the animals that have a "backbone" or spinal column composed of a series of bones known as *vertebræ*. To this group belong fishes, frogs, turtles, birds, rabbits, and human beings, for all of them have a spinal column made up of vertebræ. Insects, earthworms, and oysters, on the other hand, have no backbone; hence, they are called invertebrates (*i.e.* animals without vertebræ).

147. Summary of the classification of the invertebrates. — While the vertebrates, on account of their size, are more familiar to most people, in reality there are a great many more kinds of invertebrates than vertebrates. For example, over 300,000 different species of insects have been described, more than all other species of animals put together. The invertebrates are divided by zoölogists into ten or more *branches* or *subkingdoms*, some of the most common of which are named in the table on pages 192 and 193.

148. Summary of the classification of the vertebrates. — The *vertebrate branch* of the animal kingdom is divided into five distinct *classes*. The striking characteristics of each of these classes will be seen by studying the table on page 194.

149. Reproduction among the vertebrates. — Among the animals belonging to the two lowest vertebrate groups, namely, the fish and amphibia, the female forms eggs within the body and deposits them in the water. Before these eggs can develop, however, they must be fertilized by sperm-cells produced by the male, and this is likewise true of all the higher animals and plants. The fertilized eggs develop into embryos by the process of cell division, and enough food is stored in the egg to supply the young animal until it can secure its own food. Much the same is true also in the case of reptiles, except that the eggs are usually laid in the sand and left to develop by the warmth of the sun. There are, however, certain exceptions to the general statements made above. Some of the sharks, for example, and certain of the snakes, instead of depositing eggs that develop into embryos in the water or on land, retain the eggs, and the young are born in a form much like that of the adult.

Very few of the animals belonging to the classes that we have been discussing (namely, the fishes, amphibia, and reptiles) ever take any care of their young. The great majority of birds, however, not only build nests in which to lay their eggs, but they also brood over their eggs until they are hatched, and then the parents feed the young until they are ready to fly.

A few of the lowest mammals, like most of the vertebrates named above, lay eggs. By all the common mammals, however, the eggs are not laid, but as was the case with certain sharks and snakes, the eggs develop into a form resembling the parent, before being born. All mammals at birth, unlike birds, are unable to eat the food that is used by their parents. Hence, a form of food that is easily digestible must be furnished. This is secreted by certain cells of the adults in the form of milk. The masses of cells that secrete milk are known as *mammary glands*, and because of the presence of these glands in all animals of this the highest group of vertebrates, this class is known as the *mammals*.

Name of Branch	Example	General Characteristics	Method of Breathing	Method of Feeding	Method of Locomotion
Protozoa (first animals)	Paramecium Amœba	This group includes all the single-celled animals	Through outer surface of cell	By means of cilia or pseudopods	By means of cilia or by the movement of all the protoplasm
Porifera (pore-bearing)	Common bath sponge	Pores all over the body — many-celled, but without digestive, circulatory, or nervous systems	By means of cilia currents of water bring oxygen to all the cells and remove wastes	By means of cilia currents of water bring food to all the cells	In the mature form, sponges are fixed
Cœlenterata (hollow digestive tract)	Corals Jellyfishes	Body cavity and digestive cavity one and the same	Cells of exterior and interior bathed by water	By means of tentacles supplied with stinging cells which bring food to mouth	Corals are fixed in mature stage Jellyfish carry on locomotion by enlarging and contracting the bell-shaped body

ADDITIONAL ANIMAL STUDIES 193

Annelida (body made of rings)	Earthworm	Parts of the elongated body arranged in rngs. No jointed appendages	By means of soft, moist skin	By means of sucking mouth at anterior end	Earthworms carry on locomotion by elongating and contracting the segments, aided by bristles
Mollusca (soft-bodied)	Oyster Clam Snail	Soft body, usually covered by shell	By means of gills, or soft, moist skin	By means of cilia in clams that bring food into mouth; by raspingtongue in snails	Oysters are fixed in adult stage Clams and snails carry on locomotion by means of muscular "foot"
Arthropoda (jointed feet)	Insects Crustacea Spders	Segmented body, jointed appendages	By means of air tubes in insects; gills, in crustacea.	By means of jointed mouth parts	By means of legs and wings in insects By means of abdomen and swimmers in crayfish

Class	Example	Covering of Body	Warm or Cold Blooded [1]	Appendages used in Locomotion	Organs of Respiration
Fishes	Codfish	Scaly skin	Cold blooded	Fins	Gills
Amphibia	Frogs	Naked skin	Cold blooded	Two pairs append. Toes without claws	Gills in tadpole, lungs in adult
Reptiles	Turtles Lizards	Scaly skin	Cold blooded	Two pairs append. Toes with claws	Lungs throughout life
Birds	Robins Sparrows	Skin with feathers	Warm blooded	Anterior append. wings; poster. append. with claws	Lungs
Mammals	Cows Man	Skin with hair	Warm blooded	Paired append. with nails	Lungs

[1] Cold-blooded animals are those animals in which the temperature of the blood changes with the temperature of their surroundings. Warm-blooded animals, on the other hand, maintain under normal conditions an almost constant temperature. The temperature of the human body, for example, is 98.6° F., which is usually higher than the temperature of the earth, air, and water; consequently when we touch a fish or frog, the animal feels cold.

INDEX

A. B. = Animal Biology.　　H. B. = Human Biology.

Abdomen, A. B., 6, 151.
Abdominal cavity, H. B., 1, Fig. 1.
Absorption,
　in fish, A. B., 131.
　in frog, A. B., 110.
　in man, H. B., 98–101.
Accommodation of eye, H. B., 163.
Acetanilid, H. B., 78, Fig. 24.
Adenoids, H. B., 134.
Agar, nutrient, for growth of bacteria, H. B., 14.
Air sacs, H. B., 125, 127, Fig. 41.
Alcohol,
　as a possible food, H. B., 67.
　as a stimulant and narcotic, H. B., 67.
　effect of moderate amount on dogs, H. B., 68–72.
　effect on manual dexterity, H. B., 72.
　effect on mental activity, H. B., 72.
　effect on muscular activity, H. B., 72.
　effects of small and large quantities, H. B., 68.
　relation to body temperature, H. B., 143.
　relation to digestion, H. B., 104.
　relation to disease, H. B., 73.
　relation to life insurance, H. B., 73.
　relation to nervous system, H. B., 161.
Alcoholic beverages, H. B., 66–75.
Alimentary canal,
　of fish, A. B., 130, Fig. 98.
　of frog A. B., 110, Fig. 80.
　of man, H. B., 82, Fig. 26.
Amœba, A. B., 170–171.
Anæmia, H. B., 129.
Animal foods, composition of, H. B., Fig. 19.

Annelida, A. B., 179.
Anopheles mosquito, A. B., 46, 174, Fig. 32; H. B., 42.
Antenna,
　of bee, A. B., 31.
　of butterfly, A. B., 6.
　of crayfish, A. B., 151.
　of grasshopper, A. B., 22.
Anterior, A. B., 6.
Antiseptics, effect on growth of bacteria, H. B., 20.
Antitoxins, H. B., 35.
Anti-typhoid vaccine, H. B., 37.
Ants, A. B., 43.
Aorta, H. B., 117.
Appendages, A. B., 6.
Appendicitis, H. B., 98.
Arm, skeleton of, A. B., Fig. 48; H. B., 146, Fig. 44.
Army sanitation, H. B., 38.
Arsenate of lead, A. B., 19, 61.
Arteries,
　of fish, A. B., 131.
　of frog, A. B., 111.
　of man, H. B., 109, 112, Fig. 35.
Artificial propagation of fish, A. B., 140, Fig. 105.
Artificial respiration, H. B., 136.
Assimilation, H. B., 6.
Astigmatism, H. B., 164.
Athletics and tobacco, H. B., 77.
Auricle,
　of fish, A. B., 132, Fig. 100.
　of frog, A. B., 111.
　of man, H. B., 110, Fig. 33.

Bacilli, H. B., Fig. 7.
Bacillus tuberculosis, H. B., 31, Fig. 14.
Bacteria, H. B., 10–43.
　as foes of man, H. B., 23–43.

198 INDEX

as friends of man, H. B., 20–22.
definition, H. B., 11, Fig. 7.
occurrence, H. B., 14–20.
Barbs of feather, A. B., 68, Fig. 50.
Bathing, H. B., 141.
Bedbug, A. B., 60, Fig. 45.
Beekeeping, history of, A. B., 33.
Bees, A. B., 31–43.
Beverages, H. B., 65–75.
Bicuspid teeth, H. B., 86.
Bile, H. B., 101.
Bile duct, H. B., 101.
Bill of birds, A. B., 64.
Biologists, lives of, H. B., 168.
Bird houses, A. B., 99, Fig. 77.
Birds, A. B., 62–100.
Bivalve A. B., 181.
Black-leaf-40, A. B., 61.
Blood,
of fish, A. B., 131.
of frog, A. B., 110.
of man, H. B., 7, 107.
Blood corpuscles,
of frog, A. B., 111.
of man, H. B., 7, Fig. 5.
Blood flow, stopping of, H. B., 120.
Blood plasma,
composition of, H. B., 107.
of frog, A. B., 111.
of man, H. B., 7.
Bobolink, A. B., 79, 90, Fig. 67.
Bobwhite, A. B., Fig. 62.
Boiling meats, H. B., 54.
vegetables, H. B., 55.
Bottle poison, A. B., 1, Fig. 2.
Boxes for insects, A. B., 3, Figs. 4, 5.
Boys,
as destroyers of birds, A. B., 92.
as protectors of birds, A. B., 98.
Brain, H. B., 155.
Branches of animal kingdom, A. B., 190.
Bread making, H. B., 55.
Breastbone, H. B., 146, Fig. 44.
Breathing, definition of, H. B., 124.
Breathing capacity of lungs, H. B., 131.
Breathing movements,
of crayfish, A. B., 154.

of fish, A. B., 134.
of frog, A. B., 102.
of grasshopper, A. B., 26.
of man, H. B., 130–131.
Breathing pores, A. B., 26, Fig. 17.
Breathing tubes, A. B., 26, Fig. 17.
Broiling meats, H. B., 54.
Bronchial tubes, H. B., 125.
Bronchitis, H. B., 135.
Brood chamber, A. B., 34, Fig. 23.
Buds of hydra, A. B., 177.
Bullhead, A. B., Fig. 101.
Bumblebee, A. B., 31–33.
Burns, treatment of, H. B., 142.
Business arguments for total abstinence, H. B., 74.
Butterflies, A. B., 1–22.

Cabbage butterfly, A. B., 14, Fig. 10.
Canine teeth, A. B., 188; H. B., 86.
Capillaries,
of fish, A. B., 131.
of frog, A. B., 103, 113, Figs. 82, 83.
of man, H. B. 109, 115–116, Fig. 36.
Carbohydrates, in human body, H. B., 44.
uses of in, H. B., 51.
Carbon dioxid,
excretion of, A. B., 114, 136; H. B., 136.
production of, in man, H. B., 122.
Carpet sweeper,
use of, H. B., 26, Fig. 11.
Cartilage, H. B., 144.
Catarrh, H. B., 134.
Caterpillar, A. B., 11, Fig. 6.
Cats, as destroyers of birds, A. B., 92.
Cause,
of diphtheria, H. B., 34.
of pneumonia, H. B., 34.
of tuberculosis, H. B., 30.
Cell body, H. B., 6.
Cells, H. B., 5, Figs. 3, 4.
definition, H. B., 6.
division of, H. B., 6, Fig. 4.
of blood, H. B., 7.
of other tissue, H. B., 9.
Cellulose, H. B., 5.
Cement of tooth, H. B., 88.

INDEX

Cephalothorax, A. B., 151.
Chalk, A. B., 173.
Chest cavity, H. B., 1, 129, Fig. 1.
Chitin, A. B., 45.
Chittenden's experiments relative to diet, H. B., 61, footnote.
Chocolate, H. B., 66.
Chrysalis of butterfly, A. B., 12, Fig. 6.
Cilia,
  in windpipe, H. B., 126, Fig. 18.
  of bacteria, H. B., 11.
  of paramecium, A. B., 165, 167.
Circulation,
  of fish, A. B., 131, Fig. 99.
  of frog, A. B., 110.
  of man, H. B., 107–121.
Clams, A. B., 185.
Classes of vertebrate animals, A. B., 190.
Classification,
  of animals, A. B., 190–194.
  of birds, A. B., 73–80.
  of invertebrates, A. B., 192–193.
  of vertebrates, A. B., 194.
Clothes moth, A. B., 19, Fig. 15.
Clothing, H. B., 143.
Clotting of blood, H. B., 108.
Coagulation of blood, H. B., 108.
Cocci, H. B., Fig. 7.
Cockroaches, A. B., 31, Fig. 18.
Cocoa, H. B., 66.
Cocoon, A. B., 13, Fig. 16.
Codfish, A. B., 144–146, Fig. 108.
Codling moth, A. B., 18, Fig. 14.
Cœlenterata, A. B., 176.
Coffee,
  effect on body, H. B., 65.
  preparation of, H. B., 65.
  use and abuse of, H. B., 65.
Cold baths, H. B., 141.
Cold-blooded animals, A. B., 194, footnote.
Colds, H. B., 135.
Collar bones, 146, Fig. 44.
Colony of bacteria, H. B., 12, Fig. 11.
Colorado potato beetle, A. B., 59, Fig. 42.
Comb building of bees, A. B., 37, Fig. 27.

Composition of the body, H. B., 44.
Conditions favorable and unfavorable for growth of bacteria, H. B., 17.
Connective tissue, H. B., 3.
Conscious activities, H. B., 158.
Conservation,
  of birds, A. B., 97–99.
  of food fishes, A. B., 147–150.
Constipation, H. B., 103.
Consumption, H. B., 30–34.
Contractile vacuole, A. B., 169.
Cooking of foods, H. B., 52–56.
Coöperation of organs of body, H. B., 155.
Corals, A. B., 178, Fig. 126.
Cornea, H. B., 163.
Corpuscles,
  of frog, A. B., 111, 113.
  of man, H. B., 7, Fig. 5.
Cost of foods, H. B., 56, Fig. 22.
Cough medicines, H. B., 78.
Crab, A. B., 161, Figs. 113, 114.
Cranium, H. B., 146.
Crayfish, A. B., 151–163.
Croton bugs, A. B., 31, Fig. 18.
Crow, A. B., 88, 90, Fig. 74.
Crown of tooth, H. B., 88, Fig. 30.
Crystalline lens, H. B., 163, Fig. 52.
Cuckoo, A. B., 85, 90, Frontispiece.
Cure of tuberculosis, H. B., 32.
Cuts, treatment of, H. B., 29.

Daily diet, H. B., 60–62.
Dandruff, H. B., 142.
Decrease in bird life, A. B., 91.
Deep-sea angler, A. B., Fig. 97.
Defective eyes, H. B., 164.
Definition of a food, H. B., 46.
Dentine, H. B., 89.
Dermis, H. B., 140.
Destruction of birds, A. B., 92–97.
  effect of, A. B., 96.
Diaphragm, H. B., 1, 131, Fig. 1.
Diet, daily, 60–62.
Digestion,
  in crayfish, A. B., 157.
  in fish, A. B., 130.
  in frog, A. B., 110.
  in man, H. B., 82–98.

INDEX

of fats, N. B., 98.
of insoluble salts, H. B., 94, 95.
of proteins, H. B., 95–96, 98.
of starch, H. B., 90, 98.
Digestive ferments,
  of fish, A. B., 131.
  of man, H. B., 84.
Digestive glands,
  of crayfish, A. B., 157.
  of fish, A. B., 130.
  of man, H. B., 83.
Digestive system, H. B., 82–98, Fig. 26.
Direct metamorphosis, A. B., 29.
Disease,
  prevention of, H. B., 102.
  safeguards of body against, H. B., 42.
Diseases of respiratory organs, H. B., 134.
Dislocations, H. B., 149.
Distal, A. B., 6.
Distribution of bacteria, H. B., 16.
Dorsal, A. B., 8.
Drone bee, A. B., 36, Fig. 24.
Drugs, H. B., 78–81.
Dusting, proper method of, H. B., 26.
Dustless dusters, H. B., 26.
Dyspepsia, H. B., 103.

Ear,
  of bird, A. B., 65.
  of man, H. B., 166–167.
Eardrum, H. B., 166.
Earthworm, A. B., 179.
Earwax, H. B., 166.
Economy, in relation to foods, H. B., 56–60.
Economic importance,
  of bees, A. B., 42.
  of birds, A. B., 83–91.
  of butterflies and moths, A. B., 13–22.
  of crustacea, A. B., 162.
  of fish, A. B., 141–144.
  of frogs and toads, A. B., 118.
  of grasshoppers, A. B., 30.
  of mammals, A. B., 189.
  of protozoa, A. B., 173.
Eel, H. B., Fig. 96.
Effects of bird destruction, A. B., 96.
Egg cell,
  of bee, A. B., 36, 40.
  of bird, A. B., 70.
  of butterfly, A. B., 11.
  of crayfish, A. B., 159.
  of fish, A. B., 137.
  of frog, A. B., 114.
  of grasshopper, A. B., 28.
Eggs,
  of bee, A. B., 36.
  of butterfly, A. B., 9, Fig. 6.
  of crayfish, A. B., 159.
  of fish, A. B., 137.
  of frog, A. B., 114, Fig. 84.
  of grasshopper, A. B., 28, Fig. 19.
  of hen, A. B., 69, Figs. 52, 54.
  of house fly, A. B., 57, Fig. 41.
  of humming bird, A. B., Fig. 56.
  of lobster, A. B., Fig. 112.
  of mosquito, A. B., 43, Figs. 31, 32.
  of ostrich, A. B., Fig. 56.
Egret, A. B., Fig. 75.
Embryo,
  of crayfish, A. B., 159.
  of fish, A. B., Fig. 103.
  of frog, A. B., 116, Fig. 86.
  of hen, A. B., 70, Figs. 52, 54.
Enamel of tooth, H. B., 88, Fig. 30.
Energy, A. B., 113, 135, 158; H. B., 122–138.
English sparrow, A. B., 88, 97.
Enlargement of chest cavity, H. B., 130.
Epidermis, H. B., 139.
Epiglottis, H. B., 126.
Eustachian tubes, H. B., 167.
Excretion,
  of amœba, A. B., 171.
  of paramecium, A. B., 166, 169.
Excurrent siphon, A. B., 184.
Exercise, importance of, H. B., 102, 120, 129, 133, 151, 160.
Exodus, A. B., 30.
Expiration, H. B., 124, 132.

INDEX

Extermination,
  of house fly, A. B., 57.
  of mosquitoes, A. B., 54, Figs. 38, 39.
External ear, H. B., 166.
Eye,
  of bee, A. B., 31.
  of bird, A. B., 62.
  of butterfly, A. B., 6.
  of crayfish, A. B., 156.
  of fish, A. B., Fig. 90.
  of frog, A. B., 104.
  of grasshopper, A. B., 22.
  of man, H. B., 162–165.

False foot, A. B., 170, 171.
Farsightedness, H. B., 164.
Fats,
  digestion of, H. B., 98.
  in human body, H. B., 44.
  uses of, H. B., 51.
Feathers, A. B., 67, Figs. 49 50.
Feeding of birds, A. B., 99.
Femur,
  of bee, A. B., 33.
  of grasshopper, A. B., 24.
Fertilization,
  in bees, A. B., 36.
  in bird, A. B., 70.
  in butterfly, A. B., 11.
  in fish, A. B., 139.
  in frog, A. B., 114.
  in grasshopper, A. B., 28.
Field work on birds, A. B., 82.
Fins,
  of goldfish, A. B., 126.
  of other fish, A. B., 125.
  of perch, A. B., 121.
Fish, A. B., 120–150.
Flamingoes, A. B., Fig. 61.
Flavors of food, relation of bacteria to, H. B., 22.
Flies, A. B., 57–59.
  and typhoid fever, H. B., 37.
Flounder, A. B., Fig. 93.
Flycatchers, A. B., 80, Fig. 69.
Food economy, H. B., 56–60.
Food getting,
  of amœba, A. B., 171.
  of bee, A. B., 32, 39.
  of butterfly, H. B., 11.
  of crayfish, A. B., 157.
  of fish, A. B., 128.
  of frog, A. B., 104, 110.
  of hydra, A. B., 177.
  of mussel, A. B., 182.
  of paramecium, A. B., 165, 168.
Foods, H. B., 44–63.
  and the blood, H. B., 108.
  and the muscles, H. B., 151.
  and the nervous system, H. B. 160.
  and the skeleton, H. B., 147.
Food substances in human body, H. B., 44.
Foot,
  of grasshopper, A. B., 24.
  of mussel, A. B., 182.
Formalin as food preservative, H. B., 24, footnote.
Fossil bird, A. B., Fig. 47.
Fractures, H. B., 148.
Fresh air, importance of, H. B., 33, 129, 151, 160.
Frogs, A. B., 101–119.
Frying, H. B., 53.
Function of organ, H. B., 2.
Furnace heat, H. B., 138.

Gall bladder, H. B., 101, Fig. 2.
Gastric glands, H. B., 93, Fig. 31.
Gastric juice, H. B., 93.
Germs, H. B., 23, footnote.
Gill arch, A. B., 133.
Gill bailer, A. B., 155.
Gill chamber, A. B., 153.
Gill clefts, A. B., 132.
Gill cover,
  of gold fish, 122.
  of perch, 121.
Gill filaments,
  of crayfish, A. B., 154.
  of fish, A. B., 133.
Gill rakers, A. B., 133.
Gills,
  of crayfish, A. B., 153, 158, Fig. 111.
  of fish, A. B., 132–134.
  of mussel, A. B., 183.

INDEX

of tadpole, A. B., 116.
Gill teeth, A. B., 133.
Girls, as protectors of birds, A. B., 99.
Glands,
 of intestine, H. B., 97.
 of mouth, H. B., 91.
 of skin, H. B., 140.
 of stomach, H. B., 95.
Glottis,
 of frog, A. B., 101.
 of man, H. B., 125.
Grasshoppers, A. B., 22–31.
Gray matter, of nervous system, H. B., 157.
Growth, necessity of foods for, H. B., 45.
Gull, A. B., Fig. 59.
Gullet,
 of fish, A. B., 130.
 of frog, A. B., 108.
 of man, H. B., 92.
 of paramecium, A. B., 165.
Gypsy moth, A. B., 16, Fig. 13.

Habit, importance of, H. B., 159.
Habits, hygienic,
 of eating, H. B., 102.
 of breathing, H. B., 132.
Habitual activities, H. B., 158.
Hair, care of, H. B., 141.
Hatchery, A. B., 141.
Hawk, A. B., 78, 87, 89, 91, Fig. 64.
Headache powders, H. B., 78.
Heart,
 of fish, A. B., 131, Fig. 100.
 of frog, A. B., 107, 111.
 of man, H. B., 1, 109.
Heat, production of, in man, H. B., 122.
Hemoglobin, H. B., 128.
Hen, egg of, A. B., Figs. 52, 54, 56.
Heron, A. B., Fig. 60.
Herring, A. B., Fig. 110.
Herring gull, A. B., Fig. 59.
Hodge, Professor C. F.,
 experiments on dogs, H. B., 68–72.
 extermination of house fly, A. B., 59.

Honey,
 making, A. B., 39.
 stomach, A. B., 39, Fig. 28.
Honeybees, A. B., 33–41.
Hoof, A. B., 188.
Horse, A. B., Fig. 137.
House fly, A. B., 57, 58, Figs. 40, 41.
House mosquito, A. B., 43.
Humming bird, egg of, A. B., Fig. 56.
Hydra, A. B., 176, Figs. 124, 125.
Hydrophobia, H. B., 41, 170.
Hygiene,
 of blood, H. B., 108.
 of circulation, H. B., 119–121.
 of digestion, H. B., 102–105.
 of eyes, H. B., 165.
 of muscles, H. B., 151.
 of nervous system, H. B., 160.
 of red corpuscles, H. B., 128.
 of respiratory organs, H. B., 132.
 of teeth, H. B., 89.

Importance of birds to man, A. B., 83–91.
Importance of proper cooking, H. B. 52.
Incisor teeth, A. B., 188; H. B., 86.
Incomplete metamorphosis, A. B., 29.
Incurrent siphon, A. B., 184.
Indigestible foods, avoidance of, H. B., 62.
Infantile paralysis, H. B., 41.
Injurious birds, A. B., 88.
Injurious effects of bacteria, H. B. 23.
Insecticides, A. B., 61.
Insect net, A. B., 1, Fig. 1.
 boxes, A. B., 3, Fig. 5.
 collections, A. B., Fig. 4.
 killing bottle, A. B., 1, Fig. 2.
 spreading board, A. B., 2, Fig. 3.
Insects, A. B., 1–61.
 additional topics, A. B., 59–61.
 bees, A. B., 31–43.
 butterflies and moths, A. B., 1–22.
 destruction of, by birds, A. B., 84.
 flies, A. B., 57–59.
 grasshoppers, A. B., 22–31.
 mosquitoes, A. B., 43–57.
 moths, A. B., 13–22.

INDEX

Insoluble salts, H. B., 94.
Inspiration, H. B., 124, 130, 131.
Intemperance, cost of, H. B., 75.
Intestinal glands, H. B., 97.
Intestine,
  of fish, A. B., 130, Fig. 98.
  of frog, A. B., 108, 110, Fig. 80.
  of man, H. B., 2, 97, Fig. 26.
Invertebrates, A. B., 190, 192–193.
Involuntary muscles, H. B., 94, 151.
Iris,
  of fish, A. B., 136.
  of man, H. B., 163.
Isinglass, A. B., 142.

Jaundice, H. B., 102.
Jellyfish, A. B., Fig. 127.
Jenner, Dr. Edward, H. B., 40.
Joint, H. B., 147, Fig. 45.
Jungle fowl, A. B., Fig. 63.

Katydids, A. B., 31.
Kerosene emulsion, A. B., 61.
Kerosene treatment for mosquitoes, A. B., 55.
Kingbird, A. B., Fig. 69.
Kingfisher, A. B., 73, Fig. 58.
Kissinger, John R., A. B., 53, Fig. 36.
Koch, Dr. Robert, H. B., 30, Fig. 13.

Labial palps,
  of bee, A. B., 32.
  of butterfly, A. B., 7.
  of grasshopper, A. B., 23.
Labrum, A. B., 23, Fig. 18.
Large intestine, H. B., 98.
  absorption in, H. B., 101.
Larva,
  of bee, A. B., 40, Fig. 29.
  of butterfly, A. B., 12, Fig. 6.
  of mosquito, A. B., Figs. 31, 32.
Larynx, H. B., 125.
Lateral line, A. B., 137.
Laws relating to bird protection, A. B., 97.
  to protection of fish, A. B., 150.
Lazear, Dr. Jesse, A. B., 51, Fig. 35.
Leg, skeleton of, A. B., Fig. 51; H. B., 146, Fig. 44.

Lepidoptera, A. B., 9.
Life history,
  of bee, A. B., 40, Fig. 29.
  of bird, A. B., 70.
  of butterfly, A. B., 10–12, Fig. 6.
  of crayfish, A. B., 159.
  of fish, A. B., 137, Fig. 103.
  of frog, A. B., 114, Fig. 84.
  of grasshopper, A. B., 28, Figs. 19, 20.
  of house fly, A. B., 57, Fig. 41.
  of house mosquito, A. B., 43, Fig. 31.
  of malaria-transmitting mosquito, A. B., 46, Fig. 32.
Life insurance and total abstinence, H. B., 73.
Linen, action of bacteria in preparation of, H. B., 22.
Liver,
  of fish, A. B., 131, Fig. 98.
  of frog, A. B., 108, 110, Fig. 80.
  of man, H. B., 101–102, Fig. 2.
Lizard, A. B., Fig. 134.
Locomotion,
  of amœba, A. B., 170, 171.
  of birds, A. B., 63, 66.
  of butterfly, A. B., 8.
  of crayfish, A. B., 151, 152.
  of earthworm, A. B., 179.
  of fish, A. B., 125–127.
  of frog, A. B., 106.
  of grasshopper, A. B., 25.
  of hydra, A. B., 177, Fig. 125.
  of mussel, A. B., 182.
  of paramecium, A. B., 166.
Loss due to insect pests, A. B., 60.
Louse, A. B., 60, Fig. 44.
Lower lip, A. B., 23, Fig. 18.
Lungs,
  of frog, A. B., 103.
  of man, H. B., 2, 127, Figs. 2, 40.

Mackerel, A. B., Fig. 95.
Malaria, A. B., 47, 174.
  parasite, A. B., 49.
  transmission of, A. B., 48; H. B., 42.
Malaria-transmitting mosquito, A. B., 46, 174, Fig. 32.

INDEX

Mammals, A. B., 187–190.
Mammary glands, A. B., 187.
Mandibles,
 of bee, A. B., 33.
 of bird, A. B., 65.
 of crayfish, A. B., 156.
 of grasshopper, A. B., 23, 27.
Mantle of mollusk, A. B., 183.
Maxilla,
 of bee, A. B., 32.
 of crayfish, A. B., 155.
 of grasshopper, A. B., 24, 27.
Maxillary palps, A. B., 23, Fig. 18.
Meats,
 composition of, H. B., Fig. 19.
 cooking of, H. B., 52–54.
Menhaden, A. B., 142.
Metamorphosis, A. B., 29.
 of frog, A. B., 116.
Microbes, H. B., 23, footnote.
Microörganisms, H. B., 10–43.
Middle ear, H. B., 167.
Migration of birds, A. B., 80.
Milk, food substances present in, H. B., 46.
Milk supplies, H. B., 38.
Milk teeth, H. B., 87.
Millinery purposes, destruction of birds for, A. B., 94.
Mineral matters, in human body, H. B., 44.
 digestion of, H. B., 95.
 uses of, H. B., 52.
Mixed diet, necessity for, H. B., 61.
Moisture, effect on growth of bacteria, H. B., 19.
Molar teeth, A. B., 188; H. B., 86.
Mollusca, A. B., 181–185.
Molting,
 of caterpillar, A. B., 12.
 of crayfish, A. B., 159.
 of grasshopper, A. B., 29.
 of mosquito, A. B., 43.
Moran, John, A. B., 52, 54.
Mosquitoes, A. B., 43–56.
 as a means of transmitting malaria, H. B., 48.
 as a means of transmitting yellow fever, H. B., 50.

Moths, characteristics of, A. B., 13.
Mouth,
 absorption in, H. B., 99.
 cavity and its function, H. B., 84–92, Fig. 27.
Muscles,
 of bird's wing, A. B., 66.
 of man, H. B., 150–154.
Muscular energy, A. B., 127, 158; H. B., 123.
Mussel, A. B., 181–184.

Nails, care of, H. B., 142.
Narcotics, definition, H. B., 64.
Nearsightedness, H. B., 164.
Necessity for foods, H. B., 45.
Neck of tooth, H. B., 89, Fig. 30.
Necturus, A. B., Fig. 89.
Nerve centers, H. B., 155.
Nerve fibers, H. B., 155.
Nerve impulses, H. B., 157.
Nervous energy, H. B., 123.
Nervous system, H. B., 154–162.
Nests,
 of birds, A. B., 72, Fig. 73.
 of stickleback, A. B., Fig. 104.
Net, insect, A. B., 1.
Nettling cells, A. B., 177.
Newt, A. B., 118.
Nose cavity, H. B., 125, Fig. 39.
Nostrils, .
 of bird, A. B., 62.
 of fish, A. B., 136.
 of frog, A. B., 101.
Nucleus, H. B., 6, Fig. 4.
Nutrients, H. B., 47, footnote.

Oil glands, H. B., 140.
Organ, definition, H. B., 2; A. B. 173.
Organs,
 of circulation, H. B., 108–118.
 of digestion, H. B., 82, 98.
 of human body, H. B., 231.
 of respiration, H. B., 125–132.
Ostrich,
 bones of leg, A. B., 68.
 bones of wing, A. B., Fig. 48.
 egg, A. B., Fig. 56.

INDEX

Ovary,
  of crayfish, A. B., 159.
  of fish, A. B., 137, Fig. 98.
  of frog, A. B., 114.
  of hen, A. B., 70, Fig. 54.
Ovipositor, A. B., 28.
Owls, A. B., 78, 87, 90, Fig. 65.
Oxidation,
  in crayfish, A. B., 158.
  in fish, A. B., 135.
  in frog, A. B., 114.
  in man, H. B., 46, 122.
Oxygen,
  distribution of, H. B., 128.
  necessity for, H. B., 123.
Oysters, A. B., 185.

Pancreas,
  of frog, A. B., 108, 110, Fig. 80.
  of man, H. B., 2, 98.
Papillæ of tongue, H. B., 85.
Paramecium, A. B., 164–170, Fig. 118.
Pasteur, Louis, H. B., frontispiece, 168–170.
Pasteurization of milk, H. B., 17, 39.
Pasteurizers, H. B., 17, Fig. 8.
Pasteur treatment, for hydrophobia, H. B., 41.
Patent medicines, H. B., 78–81, Figs. 24, 25.
Pelican, A. B., 73, Fig. 57.
Peptone, H. B., 96.
Perch, A. B., 121, Fig. 90.
Perching birds, A. B., 79.
Peritoneum, H. B., 97
Peritonitis, H. B., 97.
Permanent residents, A. B., 80.
Permanent teeth, H. B., 88.
Perspiratory glands, H. B., 140, Fig. 43.
Petri dishes, H. B., 14, Fig. 11.
Phœbe, A. B., 90.
Pipefish, A. B., Fig. 92.
Plants, manufacture of food by, H. B., 51.
Plasma, H. B., 7.
  composition of, H. B., 107.
Pleura, H. B., 129.
Pneumonia, H. B., 34, 135.

Poison bottle, A. B., 1.
Pollen basket, A. B., 33, Fig. 26.
Porifera, A. B., 175.
Posterior, A. B., 6.
Preservation of food, H. B., 23.
Pressure, effect of, on bones, H. B., 148, Fig. 46.
Prevention,
  of diphtheria, H. B., 36.
  of tuberculosis, H. B., 32.
  of typhoid fever, H. B., 37.
Proboscis, A. B., 6, 10.
Production of energy, necessity of foods for, H. B., 45.
Proper posture, H. B., 153, Figs. 48, 49.
Propolis, A. B., 40.
Protective resemblance,
  of crayfish, A. B., 157.
  of toad, A. B., Fig. 87.
  of walking stick, A. B., 31.
Proteins, in human body, H. B., 44.
  digestion of, H. B., 95–96, 98.
  uses of, 51.
Protoplasm, H. B., 5.
Protozoa, A. B., 172.
Proximal, A. B., 6.
Pseudopods, A. B., 170, 171.
Pulp cavity, H. B., 89.
Pulse, H. B., 112, 114.
Pupa,
  of bee, A. B., 41, Fig. 29.
  of butterfly, A. B., 12, Fig. 6.
  of mosquito, A. B., 44, Figs. 31–32.
Pupil of eye,
  of bird, A. B., 62.
  of man, H. B., 163.
Purchase of foods, economy in, H. B., 58.
Pure Food and Drug Law, H. B., 24, 81.
Pus, H. B., 29.
Pylorus, H. B., 93.

Quail, A. B., 90, Fig. 62.
Queen-bee, A. B., 35, Fig. 24.

Rats and mice destroyed by birds, A. B., 87, Fig. 64.

INDEX

Reasons for cooking animal foods, H. B., 52.
   for cooking vegetables, H. B., 54.
Red corpuscles,
   of frog, H. B., 111, 113.
   of man, H. B., 8, 128, Fig. 5.
Reed, Dr. Walter, A. B., 50, Fig. 34.
Reflex activities, H. B., 158.
Regions of body, H. B., 1.
Relatives,
   of bees, A. B., 43.
   of grasshoppers, A. B., 31.
Repair, necessity of food for, H. B., 45.
Reproduction,
   of amœba, A. B., Fig. 172.
   of bacteria, H. B., 12, Fig. 7.
   of bee, A. B., 36.
   of bird, A. B., 70.
   of butterfly, A. B., 11
   of crayfish, A. B., 159.
   of fish, A. B., 37.
   of frog, A. B., 114.
   of grasshopper, A. B., 28.
   of house fly, A. B., 57.
   of mammals, A. B., 190.
   of mosquito, A. B., 43.
   of paramecium, A. B., 169.
   of reptiles, A. B., 185.
Reptiles, A. B., 185–187.
Respiration,
   of crayfish, A. B., 158.
   of fish, A. B., 135.
   of frog, A. B., 113.
   of man, A. B., 122–138.
   of paramecium, A. B., 168.
Retina, H. B., 163.
Review,
   of digestion, H. B., 105–106.
   of foods, H. B., 62–63.
Ribs, H. B., 146, Fig. 44.
Roasting meats, H. B., 54.
Robin, A. B., 85, 90, Fig. 71.
   eggs, A. B., 73.
Roe, A. B., 142.
Root of tooth, H. B., 88, Fig. 30.

Safeguards of body against disease, H. B., 42.
Saliva, H. B., 90–92.

Salivary glands, H. B., 91.
Salmon, A. B., 142–144, Fig. 107.
San José scale, A. B., 59, Fig. 43.
Scales of butterfly, A. B., 9, Figs. 7, 8.
Scarlet fever, H. B., 41.
Scavengers, birds as, A. B., 87.
Schleiden and Schwann, H. B., 5.
Scratching birds, A. B., 77, Figs. 62, 63.
Sea horse, A. B., Fig. 91.
Segments, A. B., 9.
Sensations,
   of sight, H. B., 164.
   of sound, H. B., 167.
Serum with antitoxin, H. B., 35.
Shad, A. B., Fig. 109.
Shaft of feather, A. B., 67, Fig. 50.
Shoulder blades, H. B., 146, Fig. 44.
Shower baths, H. B., 141.
Shrimp, A. B., Fig. 116.
Silkworms, A. B., 20, Fig. 16.
Siphons of mollusk, A. B., 184.
Skeleton, H. B., 144–150, Fig. 44.
Skeleton of arm of man and wing of ostrich, A. B., Fig. 48.
Skeleton of leg of man and of ostrich, A. B., 68, Fig. 51.
Skin, H. B., 139–143.
Skull, H. B., 146, Fig. 44.
Sleep, importance of, H. B., 129, 151, 160.
Sleeping sickness, A. B., 174.
Small intestine, H. B., 97.
   absorption in, H. B., 100.
Smallpox, H. B., 40.
Snail, A. B., Fig. 133.
Snake, A. B., Fig. 135.
Soda water, H. B., 66.
Soil fertility, relation of bacteria to, H. B., 20.
Soluble mineral matters, H. B., 95.
Soothing sirups, H. B., 78.
Soups, preparation of, H. B., 53.
Sow bug, A. B., Fig. 115.
Sparrow, A. B., 80, 90, Fig. 70.
Spermary, A. B., 70.
   of crayfish, A. B., 159.
   of fish, A. B., 137.
   of frog, A. B., 114.

INDEX

Sperm-cell,
  of bee, A. B., 36.
  of bird, A. B., 70, Fig. 53.
  of butterfly, A. B., 11.
  of crayfish, A. B., 159.
  of fish, A. B., 137.
  of frog, A. B., 114.
  of grasshopper, A. B., 28.
Spinal cord, H. B., 155.
Spiracles, A. B., 26.
Spirilla, H. B., Fig. 7.
Sponges, A. B., 175, Fig. 123.
Spore formation of bacteria, H. B., 13, Fig. 7.
Sprains, H. B., 149.
Spreading board, A. B., 2, Fig. 3.
Starch, digestion of, H. B., 90, 98.
Stegomyia, mosquito, A. B., 50, 174; H. B., 42.
Stewing, H. B., 53.
Stimulants and narcotics, H. B., 64–81, 104–105, 143, 161–162.
Sting ray, A. B., Fig. 94.
Stomach,
  of bee, A. B., 39, Fig. 28.
  of fish, A. B., 130, Fig. 98.
  of frog, A. B., 108, 110, Fig. 80.
  of man, A. B., 2, 93–97, 99, Fig. 2.
Stylonychia, A. B., Fig. 117.
Sucking tube, A. B., 6, 10.
Suffocation, H. B., 135.
Sugars as part of diet, H. B., 62.
Summer residents, A. B., 81.
Supers, A. B., 34, Fig. 23.
Swarming, A. B., 41, Fig. 30.
Sweat glands, H. B., 140, Fig. 43.
Sweeping, proper methods of, H. B., 25–29, Fig. 11.
Swim bladder, A. B., 127.
Swimmeret, A. B., 153, 159.
Swimming birds, A. B., 76, Fig. 57.

Tadpole, A. B., 116, Fig. 84.
Tapeworm, A. B., Fig. 129.
Tarsus,
  of bee, A. B., 33.
  of grasshopper, A. B., 24.
Tea,
  effect on body, H. B., 65.

preparation of, H. B., 65.
use and abuse of, H. B., 65.
Teeth,
  of fish, A. B., 128.
  of frog, A. B., 104.
  of man, H. B., 85–89.
Temperature, effect on growth of bacteria, H. B., 17.
Tendons, H. B., 3.
Tentacles, A. B., 176.
Tent for consumptives, H. B., 33, Fig. 15.
Tern, A. B., 95, Figs. 49, 76.
Thigh, of grasshopper, A. B., 24.
Thorax, A. B., 6.
Throat, H. B., 92, 126, Fig. 39.
Thrush, A. B., 79, Fig. 68.
Tibia,
  of bee, A. B., 33.
  of grasshopper, A. B., 24.
Tight clothing, effect on respiration, H. B., 133, Fig. 43.
Tissues, H. B., 3.
  definition, H. B., 4, 9.
Toad, A. B., 116, Fig. 87.
Tobacco, H. B., 75–78.
Tongue,
  of bee, A. B., 32, Fig. 25.
  of frog, A. B., 104, 110, Fig. 79.
  of man, H. B., 85.
Tonsillitis, H. B., 134.
Tonsils, H. B., 134, Fig. 27.
Total abstinence and life insurance, H. B., 73.
Toxins, H. B., 29.
  of diphtheria, H. B., 34.
Trachea,
  of frog, A. B., 101.
  of grasshopper, A. B., 26, Fig. 17.
Transformation of energy, H. B., 123.
Treatment,
  of cuts, H. B., 29.
  of diphtheria, H. B., 35.
  of pneumonia, H. B., 34.
  of tuberculosis, H. B., 32.
Trichina, A. B., Fig. 130.
Tubercles of lung tissue, H. B., 32.
Tubercles on roots, H. B., 21, Fig. 9.
Tuberculosis, H. B., 30–34.

INDEX

Turkey, A. B., 78.
Turtle, A. B., 185.
Tussock moth, A. B., 14, Fig. 11.
Tympanum, H. B., 166.
Typhoid fever, H. B., 36–38.

Umbo, A. B., 181.
Upper lip, A. B., 23, Fig. 18.
Use of foods, economy in, H. B., 59.
Uses of foods, substances, H. B., 50.
Uvula, H. B., 92.

Vaccination, H. B., 40.
Vacuum cleaner, use of, H. B., 26, Fig. 11.
Valves,
 in arteries, H. B., 114.
 of heart, H. B., 112, Fig. 34.
 of mussel, A. B., 181.
Vane of feather, A. B., 67, Fig. 50.
Vegetable foods, composition of, H. B., Fig. 20.
Veins,
 of fish, A. B., 131.
 of frog, A. B., 111.
 of man, H. B., 109, 116, Fig. 37.
Ventilation, H. B., 136–138.
Ventral, A. B., 6.
Ventricle,
 of fish, A. B., 132, Fig. 100.
 of frog, A. B., 111.
 of man, H. B., 110, Fig. 33.
Vermiform appendix, H. B., 98.
Vertebræ, A. B., 190; H. B., 144.
Vertebrates, A. B., 63, 190, 194.
Villi, H. B., 100, Fig. 32.
Vocal cords, H. B., 126.
Voluntary muscles, H. B., 151.
Von Behring, H. B., 35.
Vorticella, A. B., Fig. 117.

Wading birds, A. B., 77, Fig. 61.
Walking sticks, A. B., 31, Fig. 21.
Warbler, A. B., Fig. 72.
Warm baths, H. B., 141.
Water,
 as a part of diet, H. B., 103.
 in human body, H. B., 45.
 uses of, H. B., 52.
Water supplies, H. B., 38.
Wax glands, H. B., 166.
Web-footed birds, A. B., 76, Fig. 57.
Weed seeds destroyed by birds, A. B., 85.
Whale, A. B., Fig. 136.
Whale oil soap, A. B., 61.
White corpuscles, H. B., 7, Fig. 5.
 devouring bacteria, H. B., 29, 43, Fig. 12.
White matter of nervous system, H. B., 157.
Windpipe,
 of frog, A. B., 101.
 of man, H. B., 126, Fig. 40.
Wings,
 of bee, A. B., 32.
 of bird, A. B., 65.
 of butterfly, A. B., 7.
 of grasshopper, A. B., 25.
Winter visitants, A. B., 81.
Woodpecker, A. B., 79, 90, Fig. 66.
Worker bees, A. B., 37, Figs. 24, 28.

Yellow fever, A. B., 50, 174; H. B., 42.
Yolk,
 of fish egg, A. B., 140, Fig. 103.
 of hen's egg, A. B., 71, Figs. 52, 55.

Made in the USA
Middletown, DE
11 October 2021